# STOP MOTION ANIMATION

First edition for North America and the Philippines
published in 2013 by Barron's Educational Series, Inc.

First published by RotoVision under the title
Stop Motion Animation.

Copyright © RotoVision SA, Sheridan House, 114
Western Road, Hove, East Sussex BN3 1DD, England

*All inquiries should be addressed to:*
Barron's Educational Series, Inc.
250 Wireless Boulevard
Hauppauge, New York 11788
**www.barronseduc.com**

Commisioning Editor: Isheeta Mustafi
Editor: Cath Senker
Art Director: Emily Portnoi
Art Editor: Jennifer Osborne
Technical Reader: John Ikuma, StopMotion Magazine
Design concept: Alphabetical
Design and layout: Paul Kayser, Launchpad Design
Timeline design: Jennifer Osborne
Cover design: Emily Portnoi

ISBN: 978-1-4380-0255-2

Library of Congress Control Number: 2012948370

Printed in China

9 8 7 6 5 4 3 2 1

# STOP MOTION ANIMATION

How to Make and Share Creative Videos

**MELVYN TERNAN**

BARRON'S

# CONTENTS

**INTRODUCTION** 06

## Section One
# GETTING STARTED

## Section Two
# STOP-MOTION TECHNIQUES

### BASIC PRINCIPLES
How animation works 10
Five relevant principles 11

### EQUIPMENT AND SETUP
Cameras 14
Supports for the camera 18
Lighting 20
Capture software 24
Editing software 26
Resolution 28
Aspect ratio 29
Setups 30

### SETS AND PROPS
Tools 34
Materials 36
Sets 38
Props 42

### STORYBOARDS
The importance of storyboards 46
Storyboard layout 48
Populating a storyboard 50
Storyboard timings 52

### CLAYMATION
Overview 56
Case studies
Izabela Plucinska: *Afternoon* 58
Michael Stevenson: *Pigeon Pilfer* 60
Juan Pablo Zaramella: *At the Opera* 62
Setting up a 2D Claymation 64
Shooting a 2D Claymation 66

### PUPPET ANIMATION
Overview 68
Case studies
Kevin Parry: *The Arctic Circle* 70
Antoine Kinget and Rémi Parisse:
*Pour une Pépite de Plus. . .*
*(For a Gold Nugget More)* 72
Patrick Boivin:
*Iron Man vs. Bruce Lee* 74
Building a basic puppet 76
Shooting a Puppet Animation 80

### CHALKBOARD AND WHITEBOARD ANIMATION
Overview 84
Case studies
Hilary Grist and Mike Southworth
from Collide Entertainment:
*Angels We Have Heard on High* 86
Rebecca Foster: *Johnny and June* 88
Shooting a Chalkboard Animation 90
Shooting a Whiteboard Animation 92

## CUTOUT AND SILHOUETTE ANIMATION 08

Overview 94
Case studies
 Eimhin McNamara: *The Rooster, the Crocodile, and the Night Sky* 96
 Javan Ivey: *My Paper Mind* 98
 Camille Baladi and Arnaud Roi: *Des Pop et des Up* 100
Creating a paper puppet 102
Hinging a 2D paper puppet 103
Setting up a Silhouette Animation 104
Shooting a Silhouette Animation 106

## LEGO® ANIMATION 09

Overview 108
Case studies
 Michael Hickox: *LEGO® Mini Golf* 110
 Kevin Horowitz and Jacob Schwartz: *Writer's Block* 112
 Tomas Redigh: *8-Bit Trip* 114
Building a basic LEGO® set 116
Setting up a LEGO® Animation 118
Shooting a LEGO® Animation 120

## POST-IT NOTE ANIMATION 10

Overview 122
Case studies
 Jeff Chiba Stearns: *Ode to a Post-it Note* 124
 James Sturton: *Post-it Stop-Motion Thank You* 126
 Aaron Kaminar: *If You Ever Need Someone—The Family Bones* 128
Choosing your Post-it notes and location 130
Planning a Post-it note Animation 132
Shooting a Post-it note Animation 134

## SMARTPHONE ANIMATION

Overview 136
Case studies
 Will Studd and Ed Patterson: *Dot* 138
 Will Studd and Ed Patterson: *Gulp* 140
 Kontramax: *The Mirror Cube* 142
Setting up a Smartphone Animation 144
Down shooting with an iPhone 146

# Section Three
# POSTPRODUCTION

## EDITING AND CREATING MOVIES FOR SHARING 12

Overview 152
Image sequence to movie file: OSX 154
Editing movie files in iMovie on OSX 156
Rendering movie files for upload in iMovie on OSX 158
Image sequence to movie file: Windows 160
Editing movie files in Adobe Premiere Elements in Windows 162
Rendering movie files for upload in Premiere Elements in Windows 164

Resources 166
Glossary 168
Index 170
Contributors 174
Acknowledgments 176

# INTRODUCTION

Animation is a remarkable technique. It can grab your mind when you are young, inspire you when you are old, and amaze you at all the other points in your life.

There are many disciplines of animation, including 2D Drawn, 3D CGI, Pixilation, Mixed Media, and Stop Motion. Within each of these disciplines lie multiple techniques, but out of all of these—for me—Stop Motion is one of the most tactile, hands-on, and rewarding areas of animation.

With Stop Motion, we can control and move the world around us or create our own special world from whatever materials we please. At its very basic, Stop Motion requires only a camera, some light, and an object. The rest is up to us.

This book looks at traditional areas of Stop Motion, such as clay and puppet animation, and also current techniques, such as LEGO® animation and using smartphones to create short animated films..

Each technique is demonstrated through tutorials, and there are examples from many animators throughout the book. This book will show you the basics of Stop-Motion practice and how to format your digital work correctly for show on the world stage.

## 1868
### John Barnes Linnett

Inventor of the first "flip pad," Mr. Linnet, introduces a way to produce longer animation than the Zoetrope that preceded it. It is named the Kineograph.

## 1879
### Eadweard Muybridge

Muybridge develops a system of cameras that fire in sequence as a horse runs past them. The images play back from a glass disk projected onto a screen.

## 1906
### Mr. J. Stuart Blackton

*Humorous Phases of Funny Faces* introduces a film showing chalk drawings that come to life and interact with the artist.

# 1933
## King Kong

Released in theaters, its unique mix of live-action actors with Stop-Motion dinosaurs and a giant gorilla make this movie stand out visually and technically.

MOVIE: ALICE

# 1988
## Jan Švankmajer

Combining stop motion with live action, sometimes animating the actors (known as pixilation), Švankmajer creates his first feature movie, *Alice*.

# 1979 Lotte Reiniger

Spanning sixty years, she produces more than forty-four silhouette animations created from her own hand-cut characters. Her work remains fresh and full of vitality decades after she produces her final movie, *The Rose and the Ring*.

# 2005
## Nick Park

*The Curse of the Were-Rabbit* is the first feature film starring Wallace and Gromit, popular characters from previous thirty-minute animations.

# Section One
## GETTING STARTED

Still from *Pigeon Pilfer* by Michael Stevenson

# *Basic principles*
# HOW ANIMATION WORKS

Animation exists because of how we perceive the world around us. According to a theory known as Persistence of Vision, the human brain retains an image of what it sees for a fraction of a second after the image has changed.

It is thought that Persistence of Vision explains why, when we see a series of similar images one after the other in quick succession (as in a flip pad), we don't see a lot of different images—we see movement.

An animation flip pad is a good example of how the Persistence of Vision theory works. Look at the edge of the right-hand pages of this book and you will see there are pictures drawn along the edges. Flip the pages in quick succession and you will see the drawings move. Flip them slowly and you will see a series of still images.

Stop Motion Animation relies on the same principle, except instead of a series of drawings it's a series of photographs. The subject matter is moved slightly for each new picture taken, and when the images are played back in quick succession, we have the illusion that the objects are moving without any visual aid.

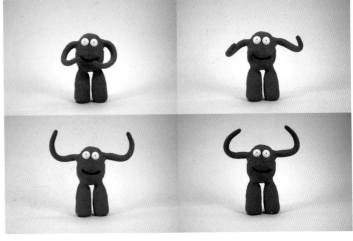

# FIVE RELEVANT PRINCIPLES

How something moves in an animation is extremely important. As humans, we have been observing the world since we were born, so we have a good idea of how things are supposed to look and move.

Over the years, a set of twelve principles about animation has been formed, originating from the book *The Illusion of Life* by Frank Thomas and Ollie Johnston. These principles were born from 2D Drawn Animation, but five of them are particularly relevant to Stop Motion Animation.

## 1. Easing in and easing out

When we move, we never "suddenly" start moving; instead we subtly "ease in" to the movements we make. When animating, it is important to observe and adhere to this, because it is one of the most fundamental principles for animation. For example, if you were to animate a clay figure waving at the camera, you would start off with small movements, progressing to larger movements as the arm goes up, and ending the wave with shorter movements again. This reads much more naturally than sudden or constant movement.

Movement with easing in and easing out–natural.

Constant movement with no easing in or easing out–unnatural.

## 2. Anticipation

First-time animators commonly make the mistake of having everything continuously moving in every frame. Our own movement is often full of pauses, so it is important that these pauses are also part of the animation. If you were to animate someone sneezing, it would be better to hold the pre-sneeze position for a couple of seconds—around 60 frames. This seems like a lot, but because the next pose is so dramatic, the hold adds to the motion of the sneeze.

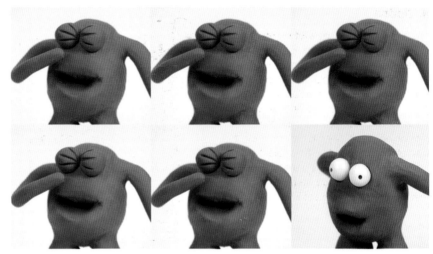

Adding pauses to movement helps it look more natural.

## 3. Staging

Staging is very important, especially within animation. Staging refers to how easy it is to see what is going on in the picture. While working on a Stop-Motion set, it is easy to view the set from any angle you please; however, it is the camera's angle that is most important because this is the angle that the audience will see when they watch your animation. To test a shot, imagine if the main subject matter were in silhouette. Could you still tell what the character was doing?

What has he got? Oh! It's a flower.

## 4. Timing

The timing principle has a lot in common with easing in and easing out. Timing relates mainly to the portrayal of weight in an animation, which sounds easy because everything you animate in Stop Motion is naturally obeying the laws of gravity. However, because you are manually moving objects bit by bit, gravity doesn't take effect as it would if that object were moving by itself. Paying attention to timing is essential if you want your subject matter to look like it exists in our world.

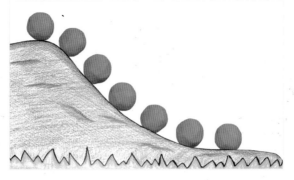

This boulder doesn't care for gravity.

This boulder obeys gravity.

## 5. Arcing

Arcing is the path that objects move along, whether it is a boomerang, a person's head, or a dog's tail. When objects move, it is difficult for them to do so in perfect straight lines unless they are mechanical devices such as pistons or trains on tracks. Keeping a moving object on a natural arc when animating helps that movement come across more naturally, as will making sure each movement is in the same direction until the object slows to a stop to change its direction.

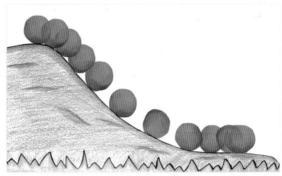

This boomerang has an incorrect arc.

This boomerang follows a naturally curved arc.

## *Equipment and setup*
# CAMERAS

Animation was originally shot on film that had to be processed before the shot could be seen. Today, digital cameras make things a lot easier because we can see what we have animated during and straight after the shot is finished. There is a wide range of cameras to choose from, but not all digital cameras are perfect for use in animation.

### Web cameras (webcams)

Generic or "no brand" web cameras are great for experimenting with animation and are relatively inexpensive. Unfortunately, they generally have a low picture resolution of 640x480 and image quality can be hit or miss.

Consoles such as the PlayStation 3 utilize their own web cameras for certain games. These cameras have good-quality lenses and can be used on a computer by installing software such as the CL Eye Driver. The image resolution is a maximum of 640x480.

The Logitech C615 can produce images of up to 1920x1080 resolution (or Full High Definition). It also has good image quality, and some models come with a tripod mount for attaching them to a tripod to keep the webcam steady while animating.

## Compact digital cameras (CDCs)

Compact digital cameras have a decent lens and can shoot high-resolution images, resulting in higher-quality images than when using a web camera. However, very few compact digital cameras work with animation capture software, so you can't see what you are shooting as you animate. Compact digital cameras are great for "shooting blind," which means that you take the frames on the camera without the aid of "onion skinning," much like when animation used to be shot on film.

Canon compact camera.

Nikon compact camera with Large Zoom.

## Digital single lens reflex cameras (DSLRs)

More expensive than web cameras and compact digital cameras, DSLRs offer much better compatibility with capture software and, because they have changeable lenses, are more versatile. Canon and Nikon DSLRs are the most popular brands to use with animation capture software because most models come with a "Live View" mode that works with the capture software to show exactly what is happening before you take a picture. The animation software can trigger the camera remotely, so the camera is not touched during animation; this avoids camera movement in the final shot.

Canon DSLR.

Nikon DSLR.

## DSLR lenses

Having control over the type of lens on a DSLR camera makes your camera a lot more versatile. A telephoto lens allows you to zoom in further, which is handy when animating because you can have the camera further back from the subject you are animating. A wide-angle lens permits you to have the camera closer to your subject matter and still see a lot of the area around it. A variable lens is the most versatile because you can zoom in or pull back to a wider shot without having to swap lenses. DSLR lenses also give you manual control over the focus, which is important when shooting animation.

A DSLR lens detached from the camera body.

Adjusting focus on a DSLR lens.

## Cell phones

Until recent years, the cameras on cell phones weren't much more than a novelty. However, the lens quality and image resolution have increased so much that software developers have written apps that allow you to use the camera on a cell phone to shoot Stop Motion Animation. Cell phones don't have a tripod mount, so it is difficult to keep the camera still while shooting animation. The lenses on cell phones are also "fixed," which means they can't zoom in or out without using digital magnification or an optical lens attachment *(see page 136)*.

HTC cell phone.

Apple iPhone.

# Digital cameras: differences

Choosing which camera to use or buy for shooting animation can be based on either what you need it to do or how versatile it will be for shooting different types of animation. The best option is to experiment with any type of digital camera you might already own instead of buying a camera purely for the purposes of animation. Remember to always check that the camera you want to use is compatible with your software.

| Camera Type | | Connection | Minimum Resolution | Maximum Resolution | Software Support | Lens Type |
|---|---|---|---|---|---|---|
| | Web camera | USB | 640x480 | 1920x1080 | Monkey Jam<br>Helium Frog<br>Dragon Frame<br>Stop Motion Pro<br>iStopMotion | Fixed<br>Some models come with manual focus control |
| | Compact digital | USB | 2448x1632 | 4000x3000 | Dragon Frame<br>Stop Motion Pro<br>iStopMotion<br>*(depending on camera model)* | Variable zoom lens |
| | DSLR | USB | 3072x2048 | 5184x3456 | Dragon Frame<br>Stop Motion Pro<br>iStopMotion<br>*(depending on camera model)* | Variable zoom lens<br>Fixed wide-angle lens<br>*(depending on lens fitted)* |
| | Cell Phone | – | 2580x2048 | 3456x2304 | Any phone application | Fixed |

# SUPPORTS FOR THE CAMERA

An important requirement when creating an animation is to keep the camera still. A camera that moves during animation, even slightly, can cause the final shot to "jitter" and move about. It is often thought that a camera supported by a stack of books or placed on a desk is good enough; however, it will soon become clear that investing in a decent support for your camera can make all the difference to the look of your animation.

### Tabletop tripod

Even a basic tabletop tripod can make a big difference in keeping your camera steady as you shoot. This type of tripod works for most cameras but isn't ideal for use with the bigger DSLR cameras.

### Full-size tripod

A full-size tripod adds a lot of versatility to how you can position your camera. With this tripod you can move the camera left and right (pan) or up and down (tilt), and some full-size tripods even have an adjustable neck to let you raise or lower the camera.

### Phone mount

You can buy phone tripod mounts that grip your phone and can be attached to a tripod to keep them steady. These can go out of date if you upgrade your phone, but you can also build your own as described in Chapter 11 *(see page 137).*

## Camera rostrum

Most Stop Motion Animation is shot horizontally: the camera is pointed across a set or at a wall. However, there are times when you need to shoot vertically, which is known as down shooting. Most tripods can be extended quite high and then tilted down to look at the ground or a tabletop, but this never results in a "true" vertical positioning of the camera.

To mount a camera for a truly vertical down-shooting position, it is best to use a camera rostrum. This consists of a heavy base for the artwork to sit on, with an extendable neck that holds a tripod plate onto which the camera is fixed.

Camera rostrums come in various sizes. Most include a set of attached lights to help to create even lighting across the artwork that is to be animated. Camera rostrums are great for animating 2D Claymation, Cutout Silhouette, and 2D Drawn Animation. Because of how the camera is mounted, the image seen on the computer will be upside down. This can be easily rectified either in the animation software you are using or afterward, when editing your animation together. *(See the online video tutorial in Chapter 12; QR code on page 156.)*

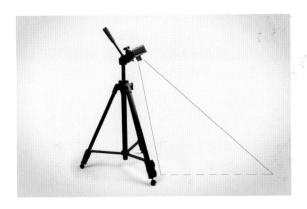

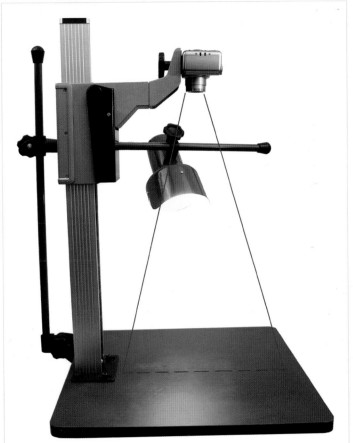

# LIGHTING

Making sure that your camera can capture what you are pointing it at is important for any type of photography as well as animation. Animations are usually shot with artificial lighting; there are many options to consider when choosing the types of light to use for your animation.

## Available light

Sometimes you have to work with any available light, which may be a mixture of outdoor and indoor lighting. When doing this, try to minimize the amount of outdoor light coming in through windows, as this can fluctuate and change. Draw the blinds, pull the curtains, or drape some dark cloth over the window to darken the room as much as possible. Light that changes during animation can result in a flickering of the image during playback.

Different types of light give off different colors. This isn't always visible to us because our brain automatically corrects the color from the different types of light we see. However, for a camera, it is important to know which types of light give off which types of color so that you can adjust the camera accordingly.

Incandescent bulb: yellow cast.

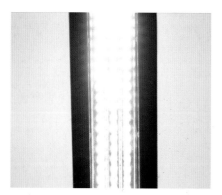

LED lights: white light.

Natural light: blue cast.

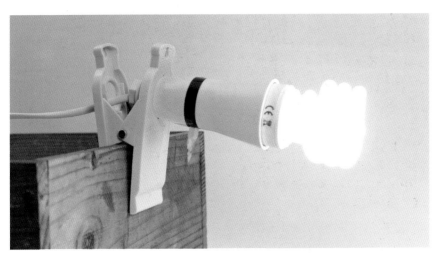

## Clip light with eco bulb

Clip lights take full-size bulbs, so they can be used with eco bulbs that give off a clean, even light and do not become hot. The clip makes it easy to attach clip lights to a heavy item, such as a wooden box or cookie jar, which you can then place near the set to help light it.

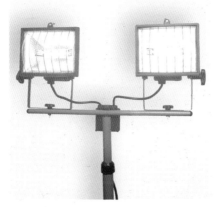

## Fluorescent desk lamp

This type of lamp is probably the most versatile and inexpensive you can use for animation lighting. The fluorescent bulb doesn't get hot and casts an even, nearly white, light. The anglepoise neck allows you to put it in almost any position, and this type of lamp often comes with either a weighted or clamp base so that it can be clamped to the edge of a table.

## Halogen lamp

Halogen lamps are generally used for lighting large sets for animation. They can become quite hot, and the bulb is usually housed behind safety glass in case the bulb explodes because of the high temperature. Some halogen bulb types can give a very strong yellow cast in their light.

## Directing lighting

As well as illuminating a set, lighting can also be used to create atmosphere or help to tell part of the story. Creative lighting is all about experimentation, so it is best to try out different techniques and see what you think of the results. However, there are a few basic lighting principles that are handy to understand.

## Diffused lighting

Some lights can be quite harsh and create strong shadow on the subject matter, especially when the light is coming from a lamp or a torch. To soften the light, you can use diffuser paper, which spreads out the light rays to create a softer light.

Diffuser paper can be bought online, or heavy-gauge tracing paper can also be used on lights that do not get hot to the touch. Always try to attach the diffuser paper so that it is slightly curved away from the light to let in air at the bulb.

## Gels

Changing the color of a light can change the feel of a shot quite dramatically and can be easily achieved by putting a colored gel in front of the light. Colored sheets of acetate also work well for this and are a little less expensive than colored gel. Again, when placing a gel in front of a light, make sure there is a gap between the bulb and the gel.

**Top to bottom, left to right:**
Harsh lighting.
Diffused lighting.
Red gel.
Blue gel.
Yellow gel.

## Bouncing light

If you find yourself in a situation where you don't have any diffuser paper and the lights you are using are too strong, you can consider bouncing the light. This technique is used to weaken the light or to direct it to a different area on the set. The most common approach is to point the light up at a white ceiling so that the light bounces and comes down as diffused light onto the set.

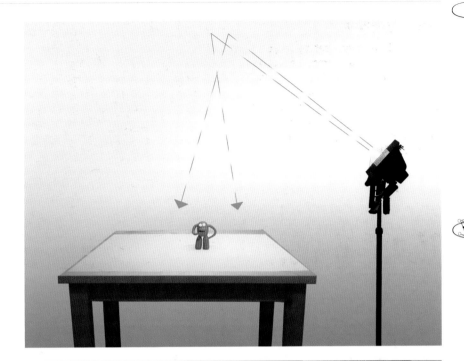

Adding a fill light refers to when the main subject matter is lit, but features such as the eyes or mouth are hard to see because they are in shadow. You can bounce a weak light off a reflector disk or a white piece of card to help bring more light to these areas. Reflector disks can be purchased online and come with a white, silver, or gold cover. The silver cover helps to intensify the bounced light, the gold cover adds warmth to the light, and the white cover diffuses the light being bounced.

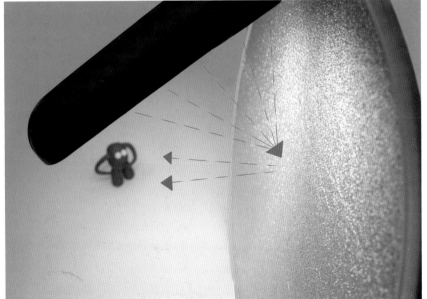

# CAPTURE SOFTWARE

Unless you are "shooting blind" using just a digital stills camera, you will need Stop-Motion capture software to capture the frames of your animation from the camera and save them to the computer. Capture software allows you to see what your camera can see, take frames remotely from the computer, and play back your animations before turning them into a movie file.

### Free software

The most popular animation capture software available for Windows computers is Helium Frog, which works with web cameras and also utilizes onion skinning.

### Inexpensive software

Bought software, such as iStopMotion for OSX computers, can be very inexpensive and often has many more functions than free software, giving you more control over your animation. This kind of software can work with some, but not all, digital stills cameras.

### Pro software

High-end Stop-Motion software brings a lot more features, such as editing, and can capture directly from DSLR cameras with full camera remote-control functionality. Software such as Dragon Frame is also used within the professional industry to shoot animated feature movies.

## How capture software works

When you download and install Stop-Motion capture software, you connect the camera or webcam, turn it on, and then launch the software. The view from your camera will either appear right away or you will need to select the camera from the menu system in the software. You can then use the controls in the software to capture the frames for your animation as demonstrated on pages 32-33.

Some software automatically saves your animation as a movie file, but it is best to set up the software to capture an image sequence. An image sequence is a folder containing all the images taken when animating. These are higher quality than a movie file and are easier to edit with later on.

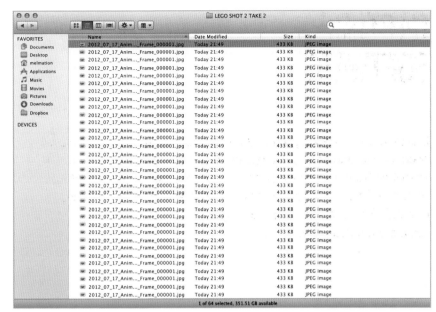

## Capture software table

| Software | System Support | Onion Skinning | Camera Support | Cost |
|---|---|---|---|---|
| Monkey Jam | Windows | No | Webcam | Free |
| iStopMotion | OSX | Yes | Webcam | $49–$499 |
| Dragon Frame | Windows/OSX | Yes | DSLR/Webcam | $295 |
| Stop Motion Pro | Windows | Yes | DSLR/Webcam | $70–$295 |
| Fingerlab iMotion HD | iOS | Yes | Built-in iSight | Free/$1.99 |
| Helium Frog | Windows | Yes | Webcam | Free |

# EDITING SOFTWARE

After capturing your animation, you could just make it into a movie file to upload to YouTube; however, it is worth taking it to the next stage and editing different shots together, adding sound, effects, and titles. There are free editing software packages available, and even some of the more expensive packages have free trials for you to try out.

### Free

iMovie is free editing software available for OSX. You can import image sequences and add sound, effects, titles, and credits as well as directly uploading your movie to YouTube.

### Inexpensive

Adobe Premiere Elements combines the most popular Pro-end tools for editing with an easy-to-use interface. Premiere Elements works on both Windows and OSX computers and can handle a lot of advance editing techniques to make the most of your animation.

### Pro

Adobe Premiere Pro offers many features that let you take your animation to the next level in terms of editing capability. While there is a steeper learning curve with Premiere Pro than with cheaper software, it may be worth investing your time in learning how to use it.

**Left:** Shot from the camera with flat colors and dark areas at the edges.

**Right:** The same shot after editing; a filter was applied to remove the dark areas.

## Postproduction

When you are ready to edit your animation together, it is a good idea to have all your image sequences ready, along with any sound or music you wish to add. This stage is called postproduction and is where you see your animation come together. Adding titles and credits to your animation can really help but it is important to keep these short and to the point. When applying effects to your animation, remember to only add effects that help to tell the story or to improve the look of your animation. For more information, see the online tutorials in Chapter 12; QR codes on pages 156 and 162.

### Edit software table

| Software | System Support | Cost | Free Trial |
|---|---|---|---|
| iMovie | OSX | Free (included) | - |
| Windows Movie Maker | Windows | Free (to download) | - |
| Adobe Premiere Elements | Windows/OSX | $99.99 | Yes |
| Adobe Premiere Pro | Windows/OSX | $799 | Yes |
| Final Cut Pro X | OSX | $199.99 | Yes |
| Sony Vegas Pro | Windows | $599.95 | Yes |

# RESOLUTION

Image resolution refers to the size of a digital image and is measured in pixels. If I were to capture images from a webcam in 720p, my image resolution would be 1280 pixels wide by 720 pixels high. If you don't tell the capture software at which size to capture the images, they may end up being a very low default resolution of 640x480. This would be a shame because all the hard work during animation might not be seen on such small pictures.

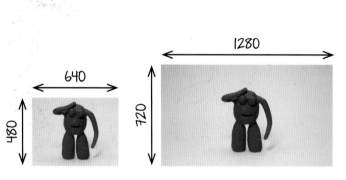

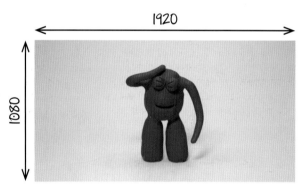

**640x480**

The smallest resolution at which most cameras capture images, 640x480, is becoming redundant as computers become faster and are able to play back longer, higher-resolution image sequences.

**1280x720**

Known as 720p, this resolution is also classified as "high definition." 720p is supported by many webcams and capture software.

**1920x1080**

Known as 1080p, this resolution is also classified as "full high definition (full HD)." The image quality of full HD looks fantastic but is only supported by high-end webcams and DSLR cameras when using Stop-Motion capture software.

# ASPECT RATIO

This refers to the shape of the picture and is important because knowing the shape of the image you are capturing determines how your animation will look when you make it into a movie file. The most common aspect ratio is 16:9 (known as "widescreen"); 4:3 is a less-used squarish shape. It is important to check the resolution and aspect ratio of your images so that the frames captured are the correct shape and size.

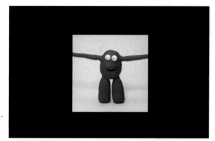

4:3 "Square".

## Frame rates

When you play a movie file, watch a DVD, or go to the cinema, you are seeing a sequence of images played back at a certain rate. Frame rates are measured per second and are determined by how many images are shown every second.

Common frame rates are 24 (for film), 25 (for European broadcast), and 29.976 (for U.S. broadcast). Make sure that your capture and edit software are set to work at the same frame rate to ensure that your animated shots play at the correct speed.

16:9 "Widescreen".

## Table of aspect ratios and resolutions

| Resolution Name | Resolution (pixels) | Aspect Ratio | System |
|---|---|---|---|
| VGA | 640x480 | 4:3 | Digital Display |
| SD | 720x480 | 4:3 | NTSC |
| SD Widescreen | 872x486 | 16:9 | NTSC |
| SD | 720x576 | 4:3 | PAL |
| SD Widescreen | 1024x576 | 16:9 | PAL |
| 720p (HD) | 1280x720 | 16:9 | NTSC/PAL |
| 1080p (Full HD) | 1920x1080 | 16:9 | NTSC/PAL |

# SETUPS

As there are many different ways you can set up equipment to create an animation, it is best to work out what you need to create the animation you want to make. You may already have a webcam or some lights, so you could see which kind of setups you can use them for. Whatever equipment you have, there are a few basics to consider when setting up to shoot Stop Motion Animation.

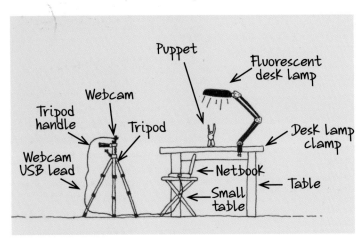

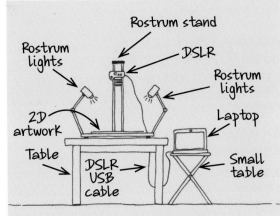

### Horizontal shooting

A common horizontal setup includes a webcam mounted on a tripod pointing at a set on a desk or a table. The webcam could be connected to a Netbook or laptop running a copy of Helium Frog and the set could be lit using a basic fluorescent desk lamp.

### Down shooting

You can use a camera on a tripod pointing down toward a table for down shooting or use a dedicated camera rostrum. In this setup, I am using a Canon DSLR attached to a laptop running Dragon Frame, with the built-in rostrum lights lighting my animation.

## 1. Shooting blind

The most basic camera setup for animation requires only a digital camera mounted on a tripod (1). This means you won't be able to use any functions found in capture software, such as onion skinning, so you will only see the results after you have finished.

## 2. Manual settings

It is important that all the settings on the camera are set to manual, including the aperture, shutter speed, and focus, so that the focus and image brightness don't change as you are animating. The setting "M" on a camera stands for "manual" and will let you control all the settings manually.

## 3. Focus

Setting the focus on any type of camera is important. Never let the camera auto focus when animating because the images will shift focus and ruin the final shot.

## 4. Timer

Set the camera timer to take a picture two seconds after you have pressed the button. This will mean the camera isn't being touched as it takes the picture.

## 5 and 6. Flash and White balance

Make sure that the flash is turned off and use the camera's white balance setting to make sure the colors are correct. Hold a white sheet in front of the camera and press the manual white balance setting (6). This will show the camera what the color white is meant to look like under the lights you are using (which will give off a color different to white), so that the camera can adjust the colors accordingly.

## Setting up Helium Frog

**1.**

When you open Helium Frog, you will be asked to choose the camera that is attached to the computer. Select the webcam you wish to use and click the arrow button.

**2.**

You will then see the resolution and frame rate settings window. Set these and click "OK" and then "Continue" to launch the interface.

**3.**

Click on "Settings" to tell Helium Frog to save the animation as an image sequence, because otherwise it will save it as a movie file. Beside the setting "Output Individual Frames" choose "Jpeg," set the quality to 100%, and click "Continue."

**4.**

Helium Frog will now save an image sequence as you shoot your animation. The image sequence is saved in a folder called "Captured Frames," which is inside the Helium Frog folder.

### Shooting on 2's

This term relates to how many frames you capture each time you move your subject matter. Shooting on 1's means you take one picture each time, whereas shooting on 2's means you take two frames each time. Most Stop Motion Animations are shot on 2's so that it only takes 15 pictures to create one second of animation when animating at 30 frames per second.

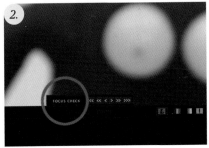

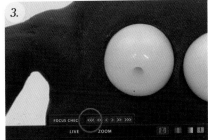

## Setting up Dragon Frame

### 1.

Once your camera is connected to the computer via USB and switched on, load Dragon Frame and choose "Create New Scene." Give your shot a name, click "OK," then choose where you want Dragon Frame to save the image files as you capture. The main interface of Dragon Frame will now load.

In the cinematography window, you can remotely set the camera's shutter speed and aperture by sliding the controls left or right.

### 2 and 3.

You can also set the focus remotely (if your camera lens is in auto mode) by clicking the focus adjusters at the bottom of the window.

### 4 and 5.

In the animation window, you can set the frame rate and aspect ratio by clicking the drop-down boxes and selecting the setting you want to use.

### 6.

Once these settings and the camera are set up, you can use the spacebar or the enter button to take a frame. Each time you do, the images will be saved to a folder in the project that you saved at the beginning.

# *Sets and props*
# TOOLS

There are many tools that are handy to have around when you are working on Stop-Motion-based animation projects. Most are DIY tools, such as drills, junior hacksaws, and steel rulers. However, there are some tools that are constantly in use throughout most Stop-Motion projects.

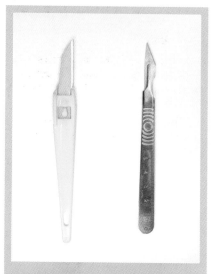

### Buy the best you can afford

A lot of time and effort is put into animation, so it is worth investing in decent tools for the job. Note that cheap scalpels can easily break, bend, or become blunt when you cut poster board. A good-quality reusable scalpel system will last longer and cut better.

### Duct tape

I can't remember a Stop Motion Animation project I have worked on where there wasn't a roll of duct tape nearby. This tape will stick to most materials (including itself) and comes in a variety of colors. It is ideal for temporarily holding things together or securing items so they don't move during animation. Never enter the studio without it!

### Hot glue gun

Continuing on the theme of adhesives, a decent hot glue gun can often save the day when creating or constructing objects for Stop Motion Animation. Glue guns produce on-demand hot glue that will quickly join most materials. In most cases, the glue can also be removed safely, making it an ideal temporary fix solution.

## C-clamps

Another staple of most animation studios is a decent set of C-clamps. These have many uses: for example, holding a set in place on a table or clamping a keyboard to the edge of a set for taking frames. C-clamps range in size, so try to have a good selection available.

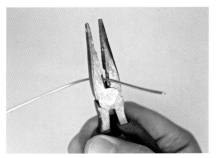

## Needle-nose pliers

When working with animation wire *(see pages 37 and 77)*, these are essential. The snipping section makes light work of trimming wire and the long plier section is very handy for removing tie-down pins on puppet feet. Make sure to buy a decent pair with a good spring action that can be operated easily with one hand.

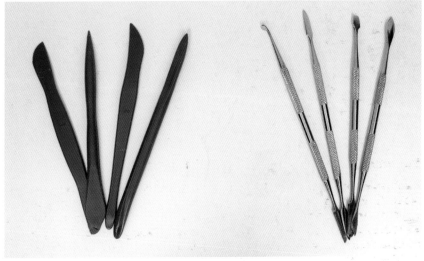

## Sculpting tools

A good set of sculpting tools will often come in handy beyond the obvious use of sculpting Plasticine or Fimo. Sculpting tools are commonly made in three different types of material: wood, plastic, and metal. Buy metal tools if you can; they will last a lifetime, whereas the wood and plastic versions can easily break.

# MATERIALS

You can use any materials you wish to create the subject matter for Stop Motion Animations as long as you can control the shape and position of the items that need to move. Like the tools mentioned for animation, there are some materials that you should try to obtain for certain projects.

### Glue safety

Always work in a well-ventilated area when working with latex or any other glue because the fumes can be overpowering.

### Epoxy glue

Epoxy glue comes in two parts that you have to mix together to activate it. Once mixed, applied, and dried, epoxy glue will stick most materials together. It is ideal for animation because a huge variety of different materials may be used to create a puppet or prop, and they often need to be stuck to one another.

### Impact glue

This type of glue is great for porous materials such as foam or cloth. Apply it to both items that you want to stick together, then leave it to become tacky before joining the parts together. Impact glue is ideal for gluing foam to a wire armature when sculpting the body of a puppet *(see Chapter 6, page 78).*

### Liquid latex

Liquid latex can be used on its own as a coating for foam or for adding to paint to give it some flexibility. Liquid latex can be bought online along with a thickening agent if you need it to be a thicker consistency.

## Foam

Foam is one of the most versatile materials for crafting animation puppet bodies. It is light and can be sculpted with a utility knife. Anything from car sponges to more dense foam found in packaging can be used. You can paint the foam or create scale costumes to put on the foam body.

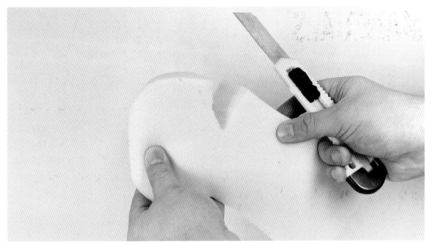

## Fimo

This clay-like material, available in many colors, can be sculpted and shaped and then baked in the oven to set hard. When the Fimo has set, it is hard enough to be sanded and even drilled into before painting. Fimo is great for making puppet heads or miniature props for your animation set.

## Aluminum wire

Aluminum wire (also known as "animation wire") is available in many different thicknesses and has the unique ability to stay exactly in the position it was bent to, whereas many other wires can move after being positioned. Aluminum wire is most commonly used for building basic animation-puppet armatures as shown on pages 76 and 77.

# SETS

Claymation or puppet-based sets can be as basic or as detailed as you want, depending on the story and theme of your animation idea. Detailed sets can mimic real-life environments, but basic sets will let the viewer focus more on the actual animation.

### Planning

Whichever type of set you need to build, it is best to take a reference from your storyboard and the initial idea for your animation. Drawing up rough plans for the dimension and size of the set is also a good starting point.

### Basic set

This style of set uses an infinity curve to give the appearance of a void in which the character is animated. A few basic props, such as shrubs, give the impression that the character is outside without having to create an entire outdoors environment.

### Detailed set

A detailed set is constructed to replicate a real-world environment, such as a living room. The props and set are constructed and painted to show exactly where the character is and what is happening within the story of the animation.

## Infinity curve

To create a set without any visible corners or walls, all you need is a large sheet of poster board. The poster board is attached to the set floor and curved up a wall and attached in place. When lit and viewed through the camera, it will give the appearance of a large, empty space within which you can place your puppet and some basic props.

You can make an infinity curve from any color of poster board to help hint at where your character is. Adding some additional basic or abstract props for reference can help show how close or far away the character is from the camera. The basic set props can be made from painted poster board and propped up with some reusable adhesive, making them quick and easy to build.

## Using basic sets

A basic set with minimal props and decoration is a great way to start off when experimenting with Puppet- or Claymation-based Animation. This type of set lets you focus on the animation and movement of the character in the set without relying on a lot of props and environmental detail to get your story across.

## Using detailed sets

Internal or external environment sets are usually constructed of wood so that they don't move while you are animating. A good choice for small sets is ¼-in. (6-mm)-thick plywood, but the bigger your set is, the thicker the wood will need to be.

### Plywood

Plywood is one of the best materials for building animation sets that are going to be used a lot. It is made of layered sheets of wood that are glued together for extra strength.

### Multi-function set

If your animation consists of different angled shots within one room, it is a good idea to build a set that can have the walls dismantled. This means you can remove certain walls and rotate the set to face the camera without having to move the camera around the set for each different shot.

### Walls

If your animation shots require seeing all three walls of a room at once, you will need to build a three-walled set. If the planned shots on your storyboard show only a background wall, you will only need to create one wall for your set.

## Security of sets

A set that is in use or "mid shot" is often referred to as a "hot set," meaning it is in use and only the animator should be working near it. Sometimes, shots aren't finished before the end of the day and so it is important that the sets, puppets, and lights are not tampered with until the animator uses them again.

An animation set usually has a camera on a tripod, lights, and a laptop positioned close together. The chance of knocking a light or bumping into the camera is quite high, so always make sure you have enough room to work.

Whenever it is possible, make sure that your set is clamped down to the desk or table you are animating on. This will help stop the set from moving while you are animating, especially when you are attaching the puppets' feet to the set.

Whether you are leaving an unfinished shot overnight or stopping animation for a few days, put a light plastic sheet over the set to stop dust building up. Never use a blanket or cotton sheet, as these can actually put dust on the set instead of keeping it off.

**Top:** A typical animation set with a camera on a tripod and lights.

**Center:** An animation set clamped to the desk to prevent unwanted movement.

**Bottom:** A plastic sheet is useful for protecting a set when it's not in use.

# PROPS

Two main types of prop for Claymation and Puppet-based Animation are practical and background props. Background props are used to decorate the environment, while practical props can be animated or used by a character within the shot.

### Made to scale

The reference for the size of your prop (and set) should be taken from the puppet you build to animate. Reference for a prop's size can be taken from the scale plan drawing of your puppet *(as shown on page 76).*

### Found props

The easiest type of prop to create for an animation is a found prop. These props can be anything from toothpaste-tube caps, used as cups, to dollhouse furniture, such as flowers, which suit the style and scale of the set you are decorating.

### Crafted props

A crafted prop is one that you build from scratch using various craft materials such as balsa wood and paint. Background props, such as trees, can be easily made from wire, DIY filler, sponge, and paint.

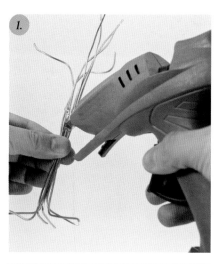

## Building a background prop: trees

Many prop items such as trees can be bought in model shops but can be quite expensive and may not be to the scale that you require. Building a tree for your set is quick and easy to do with just a few basic materials and tools.

**1.**
Cut some animation wire *(see page 37)* into strips of different lengths using a pair of pliers. Join the strips of wire together with some hot glue. Spread the bottom ends of the wire out to form the roots and the top ends of the wire out to form the branches.

**2.**
Wrap the middle of the wire strips in some sticky bandage tape to form the trunk of the tree. Continue downward onto the roots of the tree and then wrap individual pieces of wire for the root stems. Wrap the individual pieces of wire at the top for the branches of the tree.

**3.**
Using some ready-mix interior filler, coat the bandage tape to smooth over the edges. As the filler is drying, use a modeling tool to score in some lines and detail for the bark of the tree. Leave the filler to dry for the length of time specified on the filler container.

**4.**
Using acrylic paint, paint the tree and any details that you need, such as moss. To add leaves, you can either individually apply cutout paper leaves for a tree in fall or sculpt some sponge foam with a knife and paint it with acrylic paint for a tree in spring.

## Practical props

A practical prop has to be animated. It is important that practical props are built robustly so that they don't wear out or break down when being used in an animation, especially if they will appear throughout your animation.

## Umbrella

It is often worthwhile modifying an existing object to save time and effort. For example, an umbrella is a complicated mechanism and building one to scale is neither easy nor quick to do. A cocktail umbrella is a perfect starting place for this prop.

## Toaster

An item that will be seen moving in your animation, such as a toaster popping up toast, needs to be built with parts that can be animated. A detachable slider on the side along with some slots for model toast will allow the main moving parts to be animated.

## Paper

Animating paper is difficult because it naturally wants to "un-bend" every time you move it. To create paper for animation, simply paint some aluminum foil with white paint. Also try adding a drop of liquid latex to the paint to stop it from cracking when the foil bends.

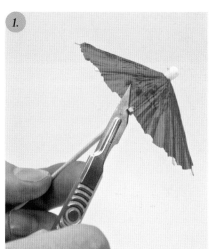

## Building a practical prop: umbrella

In this example, you will modify an everyday object to suit your need for a practical prop for animation. A cocktail umbrella replicates the mechanics and look of a genuine umbrella and can be animated bit by bit to open and close. With some modification, you can make it look like a real umbrella.

**1.**
Using a scalpel, trim off any excess pieces from the cocktail umbrella. Use some fine sandpaper to smooth down rough wood on the umbrella stem, ready for painting.

**2.**
Using black acrylic paint mixed with liquid latex, paint the paper of the umbrella both inside and out. The latex will stop the paint from cracking when the umbrella is opened and closed.

**3.**
Paint the wood and paper struts with a base coat of white acrylic, then use silver enamel paint to color in the umbrella struts and pole.

**4.**
Shape a small piece of Fimo for the handle of the umbrella and bake it to make it harden. Glue the hardened handle onto the end of the umbrella shaft using some epoxy glue.

# *Storyboards*
# THE IMPORTANCE OF STORYBOARDS

Storyboards help you to plan the different shots you want to have in your animation. They also help you plan which camera angles you might want to use, how long a shot will last, and how your shots will cut together.

Planning is essential in animation because, unlike moviemaking, everything in front of the camera has to be created and placed there. Without storyboards, a lot of time and effort would be wasted going back and redoing shots again and again.

### Inspiration for stories

Ideas for animations can be based on a story with a beginning, middle, and end. However, you don't specifically need to write a story to storyboard your animation. Taking an existing poem, favorite children's story, or even a joke you have heard can be a great basis for storyboarding an animation. Abstract animation techniques such as 2D Claymation *(see page 66)* yield better results if you just start animating and create as you go.

A common type of storyboard is one that is printed on a single sheet and fits six shots to the page. This can be handy when panning out a sequence of shots in a scene because you can see most of the shots together like in a comic book.

Storyboards also carry other information, such as the length of a shot and detail of the action within it. The aspect ratio of the final piece is also considered, with the storyboard frame drawn in a square or widescreen aspect ratio *(see page 29)*.

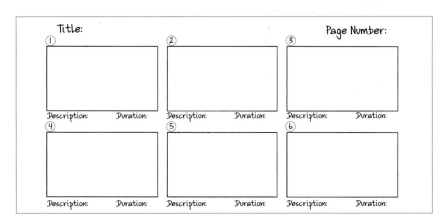

Single-sheet storyboard.

Storyboard details.

## Camera angles

Camera angles are very important; they not only help tell a story, but also assist you to plan exactly what will be seen in the shot. Close-up shots (CU) can help the audience to focus on characters when they are talking so they are not distracted by anything in the background. Wide shots (W) help to portray the environment in which the action is taking place, and medium shots (M) help to depict a closer environment. Low camera angles help to make the viewer feel intimidated, because the character on screen is portrayed as towering over the camera's point of view. High camera angles have the opposite effect; the subject matter seems vulnerable because the audience's point of view is looking down on the character.

## How many shots do you need?

It is tempting to have everything happen in just one shot when you are animating, but this can make it boring for the audience and difficult for you, especially if you make a mistake in the middle of one long shot and have to start again. Think about breaking scenes into different shots to emphasize what is happening on screen.

Low camera angles.

High camera angles.

Medium shot: Joan argues with John about doing the dishes.

Medium shot: John listens with a frown on his face.

Close-up: John wishes he were watching TV instead.

# STORYBOARD LAYOUT

There are many ways to start creating a storyboard. You can dive right in and start drawing the shots or you can write a script first so you know which shots are needed. Another way to start a storyboard is to draw images on Post-it notes or speech cards so the shots can be easily rearranged or inserted between existing shots. Once the sequence is finalized, the notes can be scanned and printed on single sheets.

Using Post-it notes to plan a storyboard helps you rearrange the scenes easily.

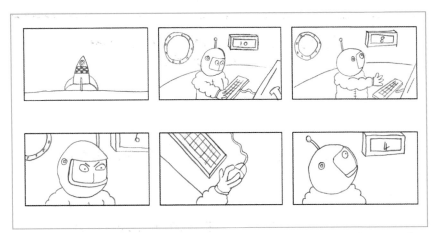

A six-shot landscape-page storyboard is handy for duplicating and sharing among a group of animators who are each animating a different shot.

A two-shot storyboard with larger frames makes it easier to include lots of detail.

## Details to include

Storyboards need to include information such as shot number, length of the shot, description of the action, and storyboard page number. This helps you to keep track of which shots have been animated and how many frames are needed for each shot.

Shot numbers can go up by one each time (e.g., 1, 2, 3), but it can be handy if they go up by ten each time (e.g., 10, 20, 30) so that if you need to insert a new shot in between other shots, you have spare numbers to use (e.g., 10, 20, 21, 25, 30).

Shot duration can be written as seconds, frames, or both. Knowing the seconds for each shot is handy for adding up the total length of the animation, while noting the number of frames a shot lasts for helps when animating it.

Including a description can help when you can't draw the action you want to happen in the storyboard frame. This space can also be used to note dialogue that is spoken during the shot.

Each page of the storyboard should have the title of the animation, along with the page number of the storyboard. This helps when you need to make a change to the storyboard or are sending a single sheet to an animator to work on.

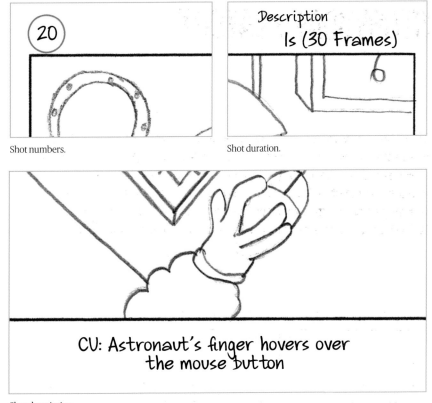

Shot numbers.

Shot duration.

Shot description.

Title and page information.

# POPULATING A STORYBOARD

Storyboards are perfect for sketching out your ideas and planning what is going to happen in your animation. The more thought you give to your planned shots at the storyboard stage, the better.

## Drawing tips

- Let ink dry before removing any pencil lines with a soft eraser.
- A soft pencil, such as a Grade 1 (3B), is perfect for shading.
- Place a small piece of clean scrap paper under your hand as you draw to avoid smudging what you have already done.
- Use pencil shading to depict lighting directions on your storyboard.

## Basic storyboards

Storyboards are usually hand drawn but don't worry if your drawing skills aren't good—even stick people can really help when planning out an animation. Don't forget to add background and prop detail to help give a sense of placement.

## Detailed storyboards

If you can draw relatively well, try adding realistic details to your storyboard. You could ink over your pencil lines to affirm what you intend to do in the animation. Adding as much detail as possible to the storyboard will help get across your intentions for the final animation.

## Adding lighting

If your animation is going to be Claymation or puppet based, include some indication as to how your sets and characters are going to be lit. This is important if you plan to use extreme lighting, such as spotlights. Basic pencil shading can work well for this.

## Adding direction

When animating, you might want to animate a camera zoom or a pan along while moving the characters in shot. For this reason, it's important to include indicators, or direction icons, to show what is moving and where.

Fat arrows can be used to indicate character movement within a shot. Thin arrows can be used to indicate camera movement. Drawing these in a different color can help to differentiate them from the drawings themselves.

TITLE: "BATTLE DINOSAUR"
DIR: M Ternam

"Men the hour has come"

"YEAH!"

"Let us charge into battle
with the man-eating dinosaurs"

"HUH?"

## Photo finish

Photo storyboards can also be created to explore in greater detail how everything might look when the animation is finished. This type of storyboard is often created when the basic sets or characters are built and you want to see how they will look in front of the camera. Photo storyboards provide an even clearer idea of how the finished animation will look.

# STORYBOARD TIMINGS

A storyboard shot could last a few seconds or nearly a minute in length. If you try to imagine how long a shot might last, I can guarantee that you will be way off the length it needs to be. It is best to act it out, especially if there is dialogue and definitely if there is only action. Here are some ways to achieve your timings.

### Stopwatch

Using a stopwatch to time yourself is one way to start, starting and stopping the clock while acting out each shot. Getting someone to time you as you carry out the actions and dialogue can also help.

### Filming

Another way to record the correct length of a shot is by filming yourself acting it out. You can then watch the footage and write down how long each shot lasted.

### Recording your voice for dialogue

Shots with dialogue can be recorded on your phone or computer using a microphone so you get a good idea of how long it takes to say the words. This method is useful for animation with lots of speaking.

## Shot length and thinking ahead

Once the basic timing is noted, it is a good idea to add edit handles of a second or two onto each shot. When you edit a cross-fade between two shots *(see Chapter 12, pages 156 and 162)*, the editing software needs to use some of the shot for the cross-fade effect. If you use a straight cut from one shot to another instead, the handles are there in case you need the shot to stay on screen a little longer than anticipated.

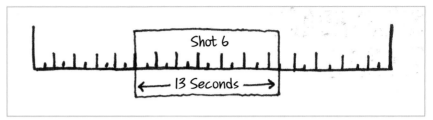

A representation of a storyboard shot length after timing it with a stopwatch.

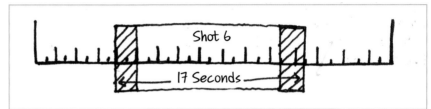

The same shot but with two extra seconds added on at either end so that there is a safe shot length ready to note on the storyboard.

To work out the length of your shots in frames, simply multiply the seconds by 24, 25, or 30, depending on the format in which you are animating *(see page 29)*. Three most commonly used frame rates and the media they are used for:

| Area of use | Film | United Kingdom & Europe | United States & Canada |
|---|---|---|---|
| Frames per second | 24 fps | 25 fps | 29.976 fps |
| Format | Film | DV PAL (UK) | NTSC (US) |

*Section Two*
# STOP-MOTION TECHNIQUES

Still from puppet-based animation sequence by Melvyn Ternan

# *Claymation*
# CLAYMATION OVERVIEW

Claymation refers to creating Stop Motion Animation using a malleable substance such as modeling clay, sculpting clay, or Plasticine. There are different types of clay that can be used for Stop Motion Animation (*see box, right*), but generally speaking, clay that doesn't dry out is best. You may need to animate it repeatedly over a period of days, so it will need to remain soft and malleable.

## Types of clay

The two most popular types of clay used in animation are:

1. Van Aken Plastalina
   Mainly used in the United States, this brand of clay is hugely popular because of its intensely bright colors and oil-based consistency, which make it ideal for animation.

2. Newplast Plasticine
   This brand of clay is mainly used for animation in Europe (for example, Aardman Animations often use this clay to model their characters). The dyes used in Newplast plasticine are less vibrant than Van Aken Plastalina colors, but the consistency works just as well for Claymation.

### Oil-based clay

Many popular Stop Motion Animations, such as *Gumby*, have been made using oil-based clay. The principle behind the animated show is the ability to change into anything—the brightly colored clay character designs add great visual appeal.

### Water-based throwing clay

Animator Kim Noce created the animated movie *After* using natural water-based clay. She added oil to the clay to stop it from drying out over the long period of time her movie took to make. Using natural clay for animation is not common, as it can easily dry out, but it does give a very unique look and texture to the finished animation.

## Claymation setup

There are two main ways in which you can shoot a Claymation movie: horizontally for set-based animation, or down shooting for 2D Layer-Based Animation.

When shooting a set-based Claymation animation, the set, lighting, and location can be as elaborate or as basic as you want. The same goes for the characters you create, although strictly speaking you should not use any material other than clay in a Claymation animation.

Using clay as a 2D medium can result in some wonderful creations and is definitely worth exploring because it provides you with the ability to sculpt, mix colors, and draw into the clay. For this setup a rostrum or a tripod on a table facing a fixed non-stick surface is ideal *(see pages 19 and 64).*

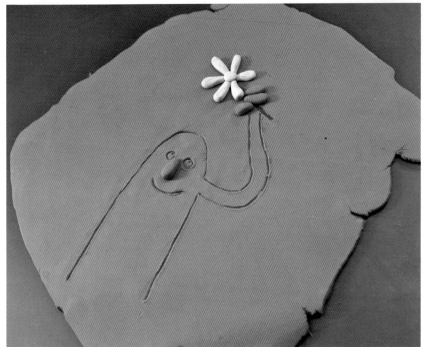

**Top:** 3D Claymation.
**Bottom:** 2D Claymation.

# Izabela Plucinska
# *AFTERNOON*

Oil-based animation clay has the unique ability to be animated to look like water, or sculpted to look like stone. However, when animating with clay to create a scene from life itself, that is when the clay becomes a form of magic.

Having made a similar animation called *Breakfast*, Izabela wanted to create another animation using the same 2D technique as her previous movie. Keeping the story simple allowed her to focus more on the actual animation itself, as she explains:

"My first idea was to use only a few elements: one room, one window, a table, man, and woman, and then to have a problem."

The clay set was on two layers: the green background is on one and the figures are animated on a sheet of glass just above so that from the camera's point of view they look like they are on top of each other.

The lights for the set were put low beside the clay with the light running across the surface to produce shadows. Everything was shot on a Canon 350D DSLR using Canon RemoteCapture software and took more than two weeks to animate.

Breaking from tradition, Izabela decided not to use a storyboard for her animation and created everything from her mind as she animated. This also let her animate at a faster than normal rate:

"For this short movie I worked without a storyboard. Ideas came directly under the camera. The black clay line I animated really fast; some days I did 10 seconds of animation."

Izabela studied art and trained in drawing, sculpture, and graphics before going to film school in Lodz, Poland, and finally studying traditional animation techniques mixed with newer, digital techniques in Germany.

"In film school, I discovered films by Rybczynski, Jan Lenica, Daniel Szczechura, and Jerzy Kucia. My professor was Piotr Dumala and I really love his work also. Polish animation has a really strong history."

While Izabela's studies have taught her many animation techniques and the discipline required for animation, it is the ability to create that fascinates her most.

**"**
You need a lot of time and patience for it, but during this time you are a kind of god, because you create a whole new world.
**"**

**Top:**
- The beautiful shadows created by the low-level lighting.

**Center (from left to right):**
- The character hides behind his newspaper.
- Backgrounds can be changed and morphed by the characters.

**Bottom (from left to right):**
- The dramatic climax caused by tension.
- Sharing the afternoon.
- With flowers.

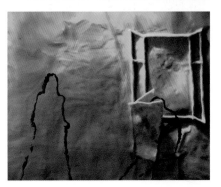

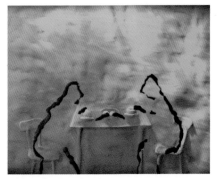

# Michael Stevenson
# *PIGEON PILFER*

Take an experience from a school field trip, a healthy appetite for Stop Motion, blocks of clay, and a small table in a cupboard. Add four months of hard work and you might just end up with a gem of a movie similar to *Pigeon Pilfer*. This short animation, with its brightly lit clay and stylized characters, offers a good example of ingenuity.

Michael always had a fascination with scale models such as maquettes, dioramas, and toy trains. These objects let loose his imagination but more so when he discovered Stop Motion Animation after experimenting with his parents' camcorder, trying to recreate the techniques seen in shows such as *Wallace and Gromit* and *The California Raisins*.

During a high-school field trip, Michael's inspiration for *Pigeon Pilfer* came from the sky above:

"Near the end of our trip we visited Pier 39, a local tourist attraction. I bought a corndog and was immediately mobbed by a flock of pigeons. They were quite bold! The experience really stuck with me."

The set was built on a small table in a storage closet so Michael had to force the perspective of his set to try to make it look larger than it actually was. Shots near the end of the animation utilized small model pigeons at the back and bigger ones at the front.

Everything in the shot is made from clay. Michael even melted large quantities of Van Aken clay in a Crock-Pot and then spread it over chicken wire to make the set pieces.

The animation was shot on a Canon DSLR hooked up to Version 1.0 of Dragon Frame, which at the time didn't support onion skinning. Michael made his own video assist by building a holder for a webcam that pointed through the eyepiece of the DSLR.

*Pigeon Pilfer* took four months to create, which included set building, lighting, and creating 120 pigeon models. A majority of that time was spent building; only the last two weeks were spent animating and editing.

" The reward is being able to look back at all the hours, the stress, the hard work, and see something rise out of it–a story, an idea, a feeling, a joke–that could not have been expressed with any less effort. "

**Top (from left to right):**
• Michael adjusting his set.
• A view of the forced perspective set.

**Center (from left to right):**
• The hapless hero.
• They don't want the food.

**Bottom (from left to right):**
• Sometimes, you just need 120 pigeons.
• Pigeons ready to pilfer.

# Juan Pablo Zaramella
# *AT THE OPERA*

This animation combines Claymation with replacement object animation to tell a short hilarious joke involving onions. The lead-up to the punch line shows various characters enjoying opera music while crying their eyes out because of its beauty . . . or something else?

Bill Plympton, along with Paul Driessen, a Dutch moviemaker and animator who primarily works in 2D Animation, inspired Juan. However, Juan also enjoyed watching Stop-Motion work created by Aardman animations and Jan Švankmajer.

"Stop Motion is a technique with a particular hand-crafted look, and I think that today a Stop-Motion animator doesn't have to try to hide visual defects in the material they use, because this is what gives Stop Motion its value."

Knowing that Stop Motion was the discipline he wanted to work in, Juan joined an animation school to study the art, covering a lot of different disciplines during his studies.

> " I was very lucky, because at the animation school they gave me a lot of freedom to experiment. I used to work in different techniques, such as sand animation, pixilation, drawing, and scratching directly over the film. "

The punch line in the animation certainly draws influence from the works of Plympton and Driessen, but the subject matter of the joke hit Juan one day when he was in his kitchen and looked at an onion.

"I saw it and I imagined it singing and people crying. The idea came to me about five years ago, but I wanted to include it as part of a bigger short about music. For around two years I left the original concept and decided to make it as a single-gag short film."

The sets were built primarily out of poster board with the puppets being mainly built from solid modeling clay. Everything was lit with two tungsten lights and filmed on a video camera using an old piece of software called Frame Thief for Mac. It took Juan two days to complete the animation and then one day to edit everything together, but this work was spread out over a period of three years.

**Top (from left to right):**
- A spectator begins to well up.
- Glycerin is used to create the tears.

**Center:**
- Hand-painted plastic beads make excellent puppet eyes.

**Bottom (from left to right):**
- Replacing the plastic tears.
- Injecting the liquid for flowing tears.
- Cutting the mouth shapes for the onions.

# SETTING UP A 2D CLAYMATION

Claymation is usually shot using a horizontal setup with clay figures on a set. But clay can also be used as a 2D medium for drawing into and creating textures. For this type of Claymation, you need to consider the type of setup you want to use and whether your animation will be abstract or character-based.

### 2D Claymation tips

- As well as drawing into the clay, try adding 2D clay props on top of your drawings to add more depth.
- Make sure your clay is thick enough to do both deep and subtle lines.
- Adding subtle fingerprints in the clay on each frame can create an animated texture.

### Down shooting

Using a camera rostrum or a tripod pointing downward is a simple way to start creating a 2D Claymation. The most important aspect of 2D Claymation is the lighting because this will help create shadows to define what the camera is seeing.

### Window shooting

If you don't have a rostrum and want to try something different, you could spread some clay out over a window that has natural light coming in through it and then score into the clay. The lines you create will be made from the light outside, which could lead to interesting results.

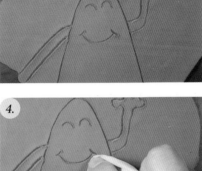

## The setup

**1.**
When you have planned out what is going to happen in the Claymation, you will need to choose the colors and the type of clay you are going to use. Use a rolling pin to spread out the clay to about 0.2 in. (5 mm) thick.

**2.**
Once you have the clay in place, use some modeling tools to draw out one of the frames of your animation. This will help you to check that the camera is in the right position and everything is in frame.

**3.**
With the test frame drawn in the clay, set up the lights to check that they are at the correct angle. The small ridges of clay along the drawn line will cast shadows, which will help to define the shape of the drawing.

**4.**
Keep some baby wipes nearby for cleaning your hands and modeling tools. Baby wipes can also be used to smooth out the clay if you accidentally add lots of unwanted fingerprints.

**5.**
Avoiding the heating vents, place some plastic wrap over your computer's keyboard, as dye from the clay can stain the keys.

# SHOOTING A 2D CLAYMATION

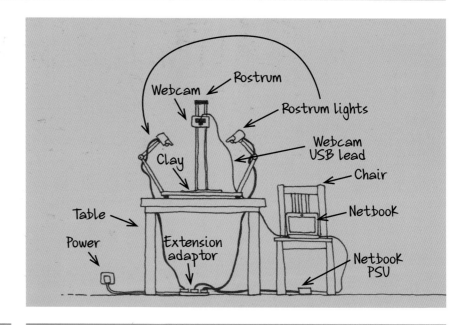

For this tutorial I am going to draw into clay to create an abstract animation. The beauty of this technique is that you can start with a simple shape and just animate free-form from there, changing the shape into something else, frame by frame. This technique can result in some truly creative animation.

## The setup

I am mounting my webcam onto a camera rostrum for a down-shooting setup. A tripod on top of a table pointing downward will also work, but a rostrum will hold the camera perfectly above the area you are animating *(see page 19)*.

Any lights can be used, but lighting that is pointed at the surface you are animating on is best because ceiling lights might cast the camera's shadow onto your work.

The webcam will be connected to a laptop running Helium Frog to capture the frames. Shooting blind with a digital camera is also an option, but not having to touch the camera while animating is ideal.

The clay will be placed onto the base of the rostrum; however, a piece of painted wood will also work. Just make sure that the piece of wood is securely attached to the tabletop to prevent it from moving.

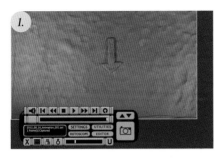

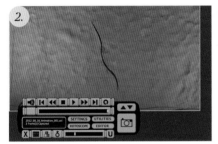

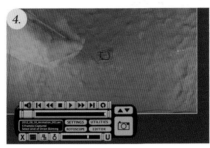

**1.**
When down shooting, the camera image will be upside down. This can be corrected afterward when editing, as discussed in Chapter 12 *(see QR code on page 156)*.

**2.**
Check that the camera is in focus by drawing a line into the clay and focusing the webcam. Also check that the line is easily seen with the lighting you are using.

**3.**
Start drawing into the plasticine with a modeling tool to create the first frame. Feel free to experiment by adding fingerprints around the drawing to create more texture.

**4.**
Use onion skinning to help to draw in the next frame and rub out any lines you don't want by smoothing the plasticine back over with your fingers.

**5.**
Change or move the lines slightly for each frame and have them change into any pattern or image that comes to mind. Try to hold on some images for ten to fifteen frames to give people time to see what you have drawn when watching the animation back.

**6.**
Experiment with introducing other colors of clay by either laying clay patterns on top of your line drawings or by smudging the other colors into the base layer of clay.

## *Puppet animation*
# PUPPET OVERVIEW

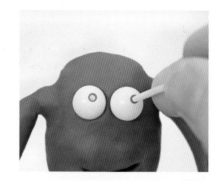

Animations such as *Wallace and Gromit*, *Coraline*, and *Paranorman* are all created using animation puppets that are made out of materials such as animation wire, foam, and specialist materials. Animation puppets are a lot of fun to animate with, but there are a few basic principles to understand before creating your own.

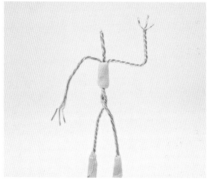

### Ball and socket

An armature is the skeleton of a Stop-Motion puppet and can be made in one of two ways: using a ball and socket system, or using animation wire. A ball and socket armature is more expensive, but its construction means that your puppet can withstand a lot of animation use.

### Wire armature

Wire-based armatures are made using strands of animation wire twisted together to form the skeleton of your puppet. This type of armature is a lot less expensive than using a ball and socket armature, but after a lot of animating the wire can become fatigued and eventually break.

### Puppet eyes and clothes

Stop-Motion puppet eyes can be made using white glass beads. The hole in the bead can be easily moved around using a toothpick or needle when animating.

Clothes can be designed in miniature for your puppet. A puppet body is usually made from soft foam to make it easy to put on the clothes, accessories, or other items.

## Standing up

Making a Stop-Motion puppet stand up on its own while animating can be difficult. To help, make sure your puppet has big feet. The larger the surface area of the puppet's feet, the easier it will be for it to stand up alone. Use a tie-down to attach the puppet's feet to the set or surface you are animating on (*see below*).

## Pin tie-down

If you have built your puppet using a wire armature and foam, you can push a pin through its feet and into the material used for the floor, such as corkboard or balsa wood. The pins can be removed using needle-nose pliers when a foot needs to leave the ground.

## Bolt tie-down

For a heavier puppet, you will need to drill holes in the set so you can put bolts up through the floor and into the feet of the puppet. This is a more secure method than using pin tie-downs but can take more effort to work with when animating.

## Magnetic tie-down

If the surface of your set isn't suitable for using pin or bolt tie-downs, you can also use Neodymium magnets (available online) to hold a metal armature in place. The feet of the puppet will need to be made of a magnetic metal for this to work.

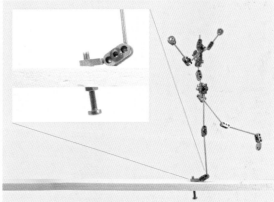

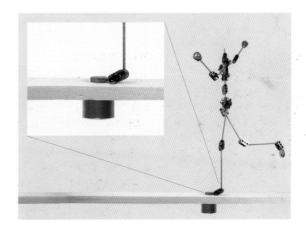

**Top:** Using a pin to make the puppet stand up.
**Center:** Using a bolt tie-down to make a heavier puppet stand up.
**Bottom:** Using a magnet to make the puppet stand up.

# Kevin Parry
## *THE ARCTIC CIRCLE*

Kevin Parry created *The Arctic Circle* as his thesis animation while studying the art form at college in Canada. A three-minute short depicting the story of greed through the character of a polar explorer, *The Arctic Circle* is both simple in its premise yet detailed in its design, animation, and storytelling.

Time constraints and limited shooting space were both influential factors when planning the animation, as Kevin explains:

"Parts of the animation, like the Arctic setting, came out of the need to create a manageable project—I didn't have to build any sets! I also knew it was to be a short film, so I tackled themes such as greed (pertaining to hunger) that would be easy to communicate to an audience."

The unique character design was arrived at following many initial sketches and drafts of how the final character should look. Kevin explains that elements such as the backpack and hood were taken from visual research and then simplified down so that the character, and therefore the actions, would read well on screen.

> 66
> The facial design came out of wanting to bring all the emotion to the eyes and keep the only follow-through (secondary animation) to the beard, which would keep the attention focused on the face.
> 99

Keeping things simple, Kevin shot the entire animation on a single 36-in. (91.5-cm) square stage, using three or four lights and an analog video camera hooked into a dedicated computer for capturing frames. To help give the appearance of vast distances, reduced-size photographs of the tree and the hut were cut out and placed in the background for certain shots.

The puppet's body was made from aluminum blocks and animation wire was used for the limbs and hands. The bulk of the body was then constructed using carved foam. The unique "elephant feet" boots worn by the explorer were designed this way to work around the method for attaching the puppet to the set.

"The character's elephant boots are a bit of a silly solution: I used 1 in. (2.5-cm) magnets to hold the puppet down, and needed to design something that would work around that cylindrical shape. I knew they were inexplicable, but that's what I really like about them."

It took over four months for Kevin to make *The Arctic Circle*, of which two and a half months were spent animating and a further month and a half were spent on postproduction. While this seems like a long time to work on a short animation, it is important to remember that the devil is in the detail, which is where the beauty of this short film lies.

**Top:**
- The explorer contemplates the meaning of the glowing box.

**Center (from left to right):**
- Design sketches for the Arctic explorer character.
- The explorer runs for the box, complete with miniature photographs in the background for scale.
- Roughing out the body shape and clothing for the explorer.
- Ready for finishing: the carved foam body and finished head ready for clothing and paint.

**Bottom:**
- Making magic: production setup for *The Arctic Circle.*

# Antoine Kinget and Rémi Parisse
## *POUR UNE PÉPITE DE PLUS...*
## *(FOR A GOLD NUGGET MORE)*

A gold miner, a desperado, a four-legged creature desperate for companionship, and a huge lump of gold. These are the ingredients for a hilarious story animated by Antoine Kinget and Rémi Parisse. Along with beautiful sets and well-animated puppets, *Pour une Pépite de Plus* demonstrates the essence of puppet-based animation.

Like most animators, Rémi and Antoine got into animation through watching cartoons and being inspired by animated films. While there is a difference between watching and creating animation, both were inspired enough by what they had seen to try it for themselves.

The idea for *Pour une Pépite de Plus* came about when Antoine and Rémi wanted to make a simple Stop Motion Animation, as Antoine explains:

> " At the beginning, we just wanted to make a little stop-motion exercise. The first idea was 'A duel between two bad guys in the desert.' And with time, we became so enthusiastic about it, that we decided to do more than just an exercise and, as great fans of Sergio Leone movies, it became a western spaghetti short movie. "

The animation was shot on a large 6 ft 6 in. by 6 ft 10-in. (2 by 3-m) table with a blue background that would be removed afterward and replaced with the sky, while polystyrene foam was used to sculpt the set pieces, such as cacti and rocks.

Ball and socket armatures were used for the human characters. The armatures were covered in a sculpted foam that was then painted and roughed up to give the characters their final appearance. The dog character was created differently, using animation wire for the armature, which was then covered and sculpted using foam.

A combination of LED spotlights and halogen lights, using yellow and blue gels to help color the desert and replicate daylight, were used to light the set. Shooting on a DSLR using Dragon Stop Motion, the animation took Antoine and Rémi ten months to create, working six to seven days a week, twelve hours a day.

Through the success of the animation, Antoine and Rémi are now freelance animators working in Paris on animation feature movies and productions.

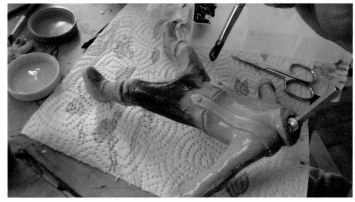

**Top (from left to right):**
• The gold digger's head and replaceable mouths.
• Painting the puppet's body.

**Center (from left to right):**
• At work animating on the set.
• A fully composited shot.

**Bottom (from left to right):**
• Basile, the four-legged critter, made using wire and foam.
• The precious gold is revealed.

# Patrick Boivin
# *IRON MAN*
# *VS. BRUCE LEE*

If you haven't got time to construct bespoke animation puppets you can always try using some action figures as your subject matter. Patrick Boivin used two iconic action figures to create a spectacular battle in Stop Motion Animation.

While Patrick has done a lot of special effects work, he has often worked in Stop Motion to create various short movies and music videos. It wasn't his first choice as a career path, but the work of other animators lured him in.

"I have never been a big fan of Stop Motion or animation of any kind, but some artists 'hooked' me because of the way their brains seemed to work."

After ordering some 12-in. (30-cm) action figures for another project, the Iron Man and Bruce Lee figures arrived in the mail first. Looking at them, Patrick came up with an idea for a short animation and, after some consideration, decided that he *had* to make it.

Using a DSLR connected to Dragon Stop Motion, Patrick spent twelve hours a day for four weeks animating on the movie. Even though this setup required some specialist equipment, Patrick believes that animation can be shot on a very simple setup:

"Today, you can do Stop Motion with your phone. There is no way you can create a complex piece of art with that kind of technology but it allows anybody to try to experiment. Then if someone feels the need to go further, he can buy a small DSLR and some great Stop-Motion software like Dragon Stop Motion (really not expensive) and become a professional."

Shooting against a green screen let Patrick animate separate parts of the puppets, and then he used software to put all the different elements back together for the final shots. As tedious as this sounds, the hard work is always worth it.

> 66
> I even had to stop for a couple of days one time because I threw a character with rage against a wall and it exploded. But when everything goes according to plan and you see the puppet coming to life, it is really rewarding . . . you feel like a magician!
> 99

**Top (from left to right):**
- Let the battle commence!
- Iron man takes out Bruce Lee.

**Center (from left to right):**
- Using a green screen to help put all the elements together afterward.
- Various tools on the set, ready at hand.

**Bottom:**
- The comedic reveal at the end of the animation.

# BUILDING A BASIC PUPPET

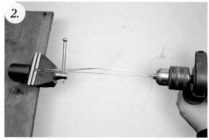

In this guide, you will build a basic Stop Motion Animation puppet that you can use as a starting point for exploring this area of animation. You will need: some animation wire, sponge foam, impact glue, bandage tape, liquid latex, and acrylic paint in various colors. You will also need some general DIY tools, including an electric drill and a hobby clamp.

**1.**
Make a full-scale drawing of your puppet on paper. This will act as a blueprint for the parts you will be making as you build the puppet, as well as allowing you to draw the layout of the wire armature that will be inside the puppet.

**2.**
Take three 24-in. (60-cm) lengths of animation wire $\frac{1}{16}$-in. (1.5-mm) thick and clamp one end of each wire into a hobby vise. Clamp the other ends into an electric drill.

**3.**
Make sure the drill is set to a low speed and turn it on to slowly twist the three strands of wire together. Do not over-tighten the wire as this will cause it to break.

**4.**

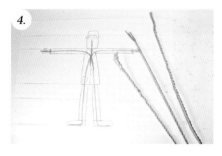

**5.**

**6.**

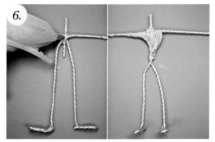

**7.**

**8.**

**9.**

**4.**
Create more twisted strands until you have a few sets of twisted wire strands made up. Make sure you have enough twisted aluminum wire to trace the layout of the armature in the drawing of your puppet.

**5.**
Bend the wire to the shape of your design, using a single length of twisted animation wire to create one arm, the body, one leg, and one foot. Bend another to make the other arm, body, leg, and foot. This layout will ensure your armature is strong and poseable.

**6.**
Use hot glue to roughly bond the wire pieces together. Twist the body of the puppet to help strengthen it. Use bandage tape to cover the armature and secure all the pieces together. The bandage tape will help give the foam body something to stick to.

**7.**
Tease apart the ends of the wire used for the arms to create the fingers for the hands. If you want your puppet to have four or five fingers, you will need to create your armature with four or five pieces of thinner animation wire in each strand.

**8.**
Cut some rough blocks of sponge foam to size for the arms, legs, body, and feet of your puppet. Make these bigger than they need to be, as you will be trimming away any excess later on.

**9.**
Cut the foam blocks in half using a long-blade utility knife. This is so that a foam block can be put on either side of the wire armature, sandwiching the armature.

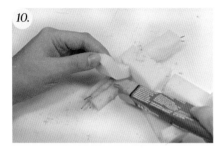

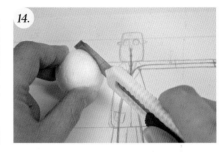

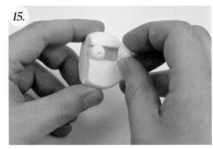

**10.**
Using impact glue, coat one half of the foam and allow it to air dry for a few minutes. Attach it to the armature. Coat the other side of the foam and attach it to the armature. Repeat this for the legs, body, arms, and feet of the puppet.

**11.**
Once the glue has dried completely, use a long-bladed utility knife to trim away the excess foam. Then use a pair of sharp scissors to add further refinement to the shape of the puppet's body.

**12.**
This puppet will have clothes that are painted on, so mix some acrylic paint with equal amounts of liquid latex in a plastic cup. Brush the mixture onto the body of the puppet, letting the latex dry completely in between coats.

**13.**
Paint the wire fingers of the puppet with some colored liquid latex. Allow the latex to dry (you can use a hair dryer on a low setting to speed up this process). Repeat this process until the puppet's fingers are twice as thick as the animation wire in the hands.

**14.**
For the head, sculpt a polystyrene ball with a craft knife to the size and shape you need. You can go back to your drawing or use your puppet body for reference to make sure the head is the correct size for the body.

**15.**
Make a recess for the eyes and fill it with some clay. Push some white glass beads into the recess to use as the eyes. The clay will help hold the glass beads in place but also allow you to move them around with a toothpick or needle when animating.

**16.**

Glue thin pieces of sponge foam to the head, using PVA glue, to build up the shape of your puppet's head. Once sculpted using a pair of scissors, coat the foam in colored latex and leave to dry.

**17.**

Attach the head to the neck of your puppet by piercing a hole in the polystyrene ball with the wire from the neck. Leave the puppet to dry completely.

**18.**

Pose your puppet into different positions to check that it moves properly before starting to animate with it.

# SHOOTING A PUPPET ANIMATION

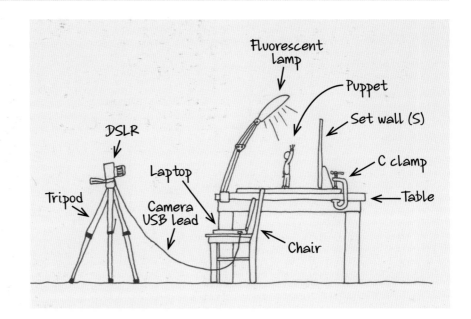

Puppet-based animation can have a lot in common with live-action filmmaking. You will need a location (a set), an actor (a puppet), lights, and a camera. As mentioned in Chapter 3 *(see page 38)*, the set for your puppet can be as simple as you wish, but if you have a particular story to tell with your animation, you will need to consider a more specific setting. Either way, the most important thing to do is experiment and have fun when animating your actor.

## The setup

For this example, I am using the basic animation puppet as created in the previous section *(see pages 76–79)*. My three-walled set can be dismantled as necessary, depending on which angle I am shooting from.

For the software, I will be using Dragon Frame to capture frames from a DSLR. The DSLR will be mounted on a tripod that will be set back from the set and zoomed in to leave me room to animate on the set.

I will be using anglepoise fluorescent lamps to light my set, with some diffusion paper to help soften their light.

**1.**
Before starting to animate, make sure that your set is securely clamped to the table or desk you are working on. Make sure that the walls of your set are secure, using wooden props and some duct tape.

**2.**
Once you have the lights set up and camera connected to the computer, launch Dragon Frame and use the on-screen controls to adjust the camera's focus and exposure *(see page 33)*. Take a few test shots to make sure that everything is set up correctly.

**3.**
Get your puppet and props into position for the first shot. It is a good idea to keep your storyboard handy, as you will need to refer to it for each shot. Here, I am making sure that the puppet is secured to the set with pin tie-downs.

**4.**
I am shooting this animation on 2's. This means I take fewer pictures for each second of animation and the movements don't have to be as refined as when shooting on 1's.

**5.**
The newspaper is made out of foil and so is easy to animate. To help make the puppet "hold" the newspaper, I am using pieces of reusable adhesive to attach the newspaper to his hands.

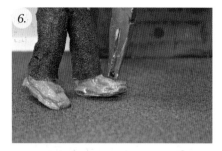

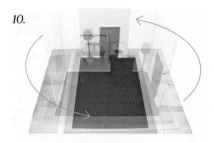

**6.**
When the puppet needs to take a step, I can remove the pin tie-down by using a pair of needle-nose pliers *(see page 35)*. The floor of the set is made out of corkboard, so the tie-downs are easy to remove.

**7.**
To make the puppet blink, I use a piece of skin-colored clay to cover the eyes for two frames and then remove them. When this is played back, it looks like the character has blinked.

**8.**
The character's mouth was added by cutting out some black poster board and attaching it to the head with a piece of reusable adhesive. To change the shape of the mouth, I switched it to a different-shaped piece of poster board between frames.

**9.**
You can use a toothpick or thick needle that can fit into the hole of the glass bead to move the eyes. Make sure that both pupils point in the same direction. Check the position of the puppet's eyes on the capture screen to make sure they are looking in the right direction.

**10.**
For the last few shots of the animation, I need to see only one wall of the set. I can take down the other walls and rotate the set so that the wall with the door in it is facing the camera.

**11.**
Check that the umbrella opens and closes easily before animating it. When the puppet picks up and holds the umbrella prop *(see page 45)*, use some reusable adhesive to help it stay in his hands.

## 12.

I dressed the area behind the door with some blue poster board, lit with a fluorescent lamp, to simulate daylight. For further effect, you could paint clouds onto the blue poster board.

## 13.

To animate wind pulling on the umbrella, simply animate the open umbrella moving up and away from the puppet while he is still holding it. Remember the "staging" principle as discussed in Chapter 1 (see page 12).

## 14.

When the puppet flies off into the air, I use my hand to lift him off the ground a little farther each frame. Make sure that your hand and arm don't cast a shadow onto the set or the puppet.

## 15.

Use the onion-skinning tool in Dragon Frame to check that you are moving the puppet the same distance each time and that the direction of the puppet doesn't change as he flies through the air.

# *Chalkboard and Whiteboard Animation*
# OVERVIEW

Chalkboard Animation has existed since 1906, when Mr. J. Stuart Blackton produced *Humorous Phases of Funny Faces*. In this animation, a person draws characters that then come to life performing basic movements. It is no surprise that a medium such as chalk was one of the first to be utilized for animation.

### Chalk

After a picture is taken of the chalkboard, the image is rubbed away so the next drawing can be made. Chalkboards lend themselves well to animation because part of the previous picture can still be seen, helping you to draw the next image in the right place.

### Dry-erase marker

A modern-day equivalent of Chalkboard Animation is the whiteboard, used with dry-erase markers. The whiteboard is used in the same way as the chalkboard, except that the marks made on a whiteboard can be cleaned off completely and more precisely.

### Made blackboard

Chalkboards can be found in toy stores or educational supply stores; however, they are often very small. You can make a blackboard by purchasing some wood at any size you wish and painting it with special blackboard paint found in most craft stores.

## Chalkboard

While whiteboard and chalkboard as media share common visual characteristics, chalkboard lends itself more to intricate, detailed drawings. Chalk can be used to create subtle shading on what you draw and also works well when blending colors together in order to create new colors.

Chalk does not produce as "clean" a mark as a dry eraser. Chalk marks can be cleaned up with items such as a damp sponge or even a moist Q-tip. Neatening up an image drawn on a chalkboard requires more effort but can result in interesting visuals.

## Whiteboard

Even though both chalkboard and whiteboard animation are strictly Drawn Animation, they can still have a Stop Motion feel to them because of the patterns and marks left behind by previous drawings. Some animators also like to include themselves in the animations by interacting with the drawings frame by frame. Whiteboard animation works quite well for this technique because it is quicker and easier to wipe away part of the previous image. This allows the animator to animate, take frames, and be in the animation all at once!

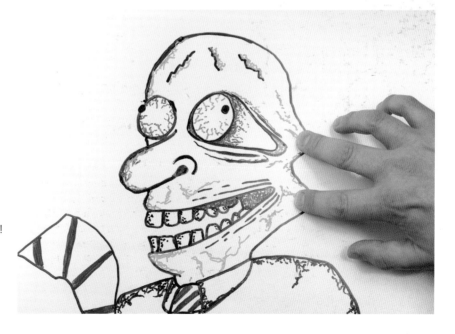

# Hilary Grist and Mike Southworth from Collide Entertainment
## *ANGELS WE HAVE HEARD ON HIGH*

Pixilation is a technique in which people are used as the animation medium, and it can create interesting results. Mixing pixilation and chalkboard animation, Mike Southworth of Collide Entertainment created an extremely creative music video for one of Hilary Grist's songs.

Hilary has always enjoyed drawing as well as recording her music. When she was growing up, she was inspired by cartoonists such as Charles Schulz, Bill Watterson, and Gary Larson. In recent years, she has been inspired by creative animation videos on YouTube and decided to try to fuse her drawing and music together.

"Stop Motion has a very 'hand-made' quality that is quirky and charming. When a video has a hand-made vibe to it, people forgive a lot of production cost. I feel like a video that is made by hand can have a lot more heart than a video that has a huge budget and computer graphics."—Hilary

Mike collaborated with Hilary on every aspect of the project, with Mike undertaking the technical side of the job, such as photography and editing, and Hilary doing the drawing and animation.

Throughout the process, Hilary and Mike bounced ideas off each other. Although Hilary had drawn up a storyboard for the animation, they added in little ideas that they came up with as they worked.

The video was created in one full day, including the animation, shooting, and editing. It was shot using Dragon Frame and then edited using Adobe Premiere Pro. The chalkboard in the animation takes up the entire frame, so this had to be built from scratch.

> **"**
> We built a set in our apartment with an 8-by-8-ft (2.4-by-2.4-m) piece of drywall, painted with chalkboard paint, and used lamps with compact fluorescent bulbs.
> **"** –Hilary

Working on her new album *Tomorrow Is a Chance to Start Over*, which will feature a fantasy picture book, Hilary has plans to create more chalkboard animations for her songs in the future.

**Top (from left to right):**
- "Outside" with happy chalk clouds.
- Cold, Hilary draws a scarf, which she plucks from the picture.

**Center (from left to right):**
- Hilary leans in to breathe on the window.
- Entering a door that Hilary has just drawn.

**Bottom (from left to right):**
- Indoors, Hilary draws a chalk mandolin.
- The mandolin then becomes real.

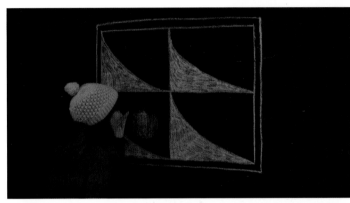

# Rebecca Foster
# *JOHNNY AND JUNE*

After being inspired by the work of graffiti and street artist Blu, Rebecca Foster decided to create her own chalkboard-based animations at home. *Johnny and June* was created for a song of the same name by the band The Hours. The creative development of the images along with striking colors against the blackboard background bring this piece a charm all of its own.

Rebecca was initially inspired about animation when she attended an experimental animation course, where she had to scratch images onto 16mm film with a pin to produce movement when the movie was played back on a projector.

"It took days but I loved the results. I had always loved animation, but this was the first time I really had a go!"

After an accident that meant Rebecca was housebound for a while, she decided to use the time to create the *Johnny and June* animation. Images in the video were inspired by the song lyrics but also by Damien Hirst, who art-directed visuals for the band.

Setting up a chalkboard in her conservatory to use natural lighting, Rebecca took three to four weeks to produce the final video, with the drawing stage taking around two weeks and requiring fifteen sticks of colored chalk to complete.

"It can be very slow and you learn as you go, but the possibilities are endless and I never fail to be excited when I see marks I have made come into their own existence."

The animation was shot on a tripod-mounted compact digital camera (Canon Ixus 80 IS), then Adobe Photoshop was used to even out the contrast on individual frames caused by the light changing from day to day. The frames were then edited together with the music using Apple's iMovie.

> **"**
> One of the amazing things about digital cameras is the fact that they open doors to anyone with a bit of patience to create their own animations at home. You need very little to get started and can watch the results back on your camera.
> **"**

**Top:**
- Vibrant chalk contrasts with the blackboard.

**Center (from left to right):**
- Two butterflies dance with each other before becoming one.
- A bird flies over the ocean.
- A Damien Hirst-inspired skull visual.

**Bottom (from left to right):**
- A bird prepares to fly past twinkling stars.
- Abstracted face and hands that merge into two birds.

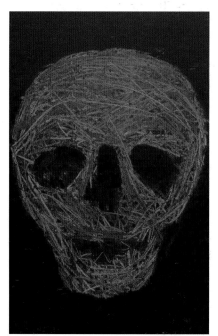

# SHOOTING A CHALKBOARD ANIMATION

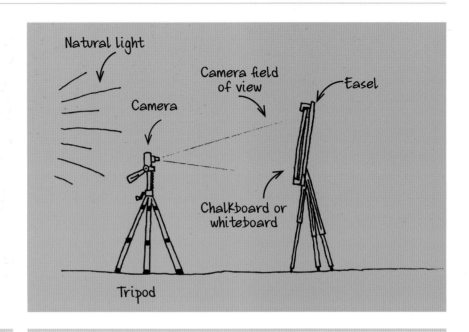

In this tutorial we use a basic setup to create an animation using nothing more than natural light, a digital camera, a blackboard, and some chalk. The beauty of creating Stop Motion Animations on a blackboard is that you have the freedom to redraw and correct the chalk line as much as you need until you are ready to take a frame. Along with a blackboard you will need a packet of white and colored chalks, a small sponge, and a soft cloth.

## The setup

For this setup I used natural light, as this helped to add character to the finished animation. You can use a light if you are working at night or want perfect control over the finished animation, but it is worth experimenting with natural light to see what effects can be achieved.

To hold the blackboard in place I used an artist's easel, which can be readily found at any art store. You can also lean the blackboard up against a wall, but this makes it less accessible and it can be harder to get the camera in position to take the frames.

Because I used an easel, I needed to make sure that the easel sections were secure and that it was firmly fastened to the floor; I did this by using small sections of duct tape on each leg. This helped reduce any shake or wobble in the easel as I was drawing and erasing the images.

The camera I used was just a basic compact digital stills camera *(see Chapter 2, page 15)*. It was mounted on a tripod and set to fully manual. Remember to keep the camera back a little and zoom in so that you have plenty of room to draw on the blackboard.

This method is referred to as "shooting blind" and is how animators used to work before digital video came along. It does not use onion skinning, as we are not recording the frames to a computer using animation software; however, blackboard animation creates its own natural onion-skinning effect, as you will see.

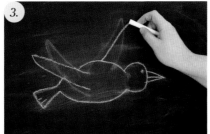

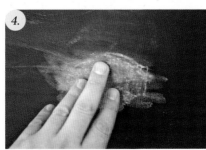

### 1.
Using the white chalk, I drew a simple side view of a bird with its wings up in the air as if flying.

### 2.
I took a frame on the camera and then used the soft cloth to erase the wings of the bird. When erasing on the blackboard don't rub too hard, as you want to leave a trace of the previous picture behind to help you redraw what you need for the next frame.

### 3.
Next, I redrew what I had erased, but this time drew the wings moving down a little. This is where you will notice the "natural onion skin" effect of drawing on a blackboard, which helps you get the next drawing in the right place.

### 4.
Repeating this process, I moved the wings of the bird up and down. I also animated some simple clouds passing the bird to help give the illusion that it was flying through the sky. I did this by using some colored chalk and smudging it slightly with my fingers.

### 5.
As you animate, you may find that the blackboard becomes quite messy with lots of previous drawing showing through. Should this happen, use a slightly damp sponge to clean up around your current frame. You can also use the corners of a damp sponge to neaten up your pictures as you go.

# SHOOTING A WHITEBOARD ANIMATION

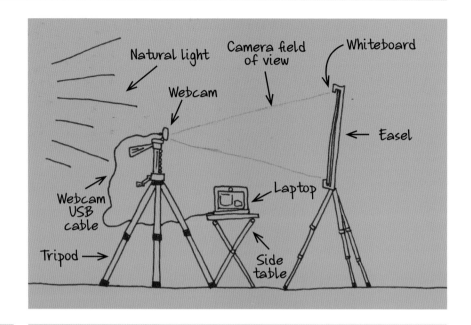

When shooting a Whiteboard Animation, follow the same fundamentals as when shooting a Chalkboard Animation. The main difference is that you use dry-erase markers with a whiteboard. If you want to include yourself interacting with the drawings, it is best to shoot using Stop-Motion software and a webcam so that you can use onion skinning to make sure your hand is in the right place each time you take a frame.

## The setup

We are using natural light to illuminate the whiteboard, as this will help add to the visual style of the finished animation.

The whiteboard is held in place using an artist's easel, which is secured in place on the floor to keep it steady. This allows clear access to the whiteboard.

The webcam (Logitech C615) is mounted on a standard camera tripod. If you are using this setup, make sure that your webcam has a tripod mount on its base to make this possible.

I am using Helium Frog, a piece of free Stop-Motion software, because it works well with webcams and also has the function of onion skinning (*see pages 24 and 32*).

**1.**

**2.**

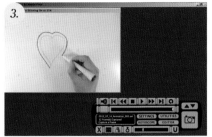

**3.**

**4.**

**5.**

**1.**
I planned out the design I wanted to animate on the whiteboard. In this example, I am going to make the design appear piece by piece, zooming toward the camera.

**2.**
Start by drawing a small dot for where the first element of the design is going to appear (in this case, the heart). Take a frame on Helium Frog, rub out the dot using a cloth, and draw a slightly bigger heart around where the dot was drawn.

**3.**
Repeat these steps, making the heart shape bigger until it is the same size as the drawing. This is where onion skinning is handy; you can check you are making the next image the right size compared to the last one.

**4.**
I repeat this process for the other elements of the design until everything matches my original drawing. I keep drawing a new heart inside the completed one over and over so it looks like the heart is beating.

**5.**
Unlike chalk, dry-erase markers will wipe off completely if you rub them with a cloth or a Q-tip. When you are using onion skinning, be careful not to wipe away too much of the drawing by accident.

# *Cutout and Silhouette Animation*
# OVERVIEW

08

Cutout and Silhouette Animation share a lot of similar visual traits to the imagery created by shadow puppets and have a very long history in animation. While Silhouette Animation looks complex, it requires nothing more than a paper puppet and a light box to create its unique look and charm.

### Lotte Reiniger

Twentieth-century German animator Lotte Reiniger created some of the most celebrated silhouette animations. Her intricate puppets and beautiful background scenery are as remarkable as the quality of the animation in her short movies.

### Hinged cutout puppets

Cutout puppets are easy to make and contain far greater detail than silhouette puppets. The material used for the puppet pieces can be brightly colored and created using an array of different craft materials.

### Silhouette puppets

While silhouette puppets lack color or drawn detail, they often have more refined and intricate shapes to help the viewer identify the characters. Silhouette puppets are usually made from stiff black poster board.

### Light box

Placing paper puppets onto a light box creates the silhouette. A light box is simply a box with a glass or Perspex lid that light shines through. Since we do not see the full detail of the puppets, we are more focused on the animation and movement of the characters, which gives Silhouette Animation its charm.

## The visual look

Cutout Animation can often have a "montage" or "graphic" feel to the visual look. This is because almost any 2D material can be used to create the imagery and puppets for a Cutout Animation. Real grass, magazine pictures, fabric, buttons, and wallpaper can all be used for this type of animation. While the puppets need to be practical and formed from easy-to-use material such as poster board, the static backgrounds can be made from any material that helps to build up the picture.

When first experimenting with Silhouette Animation, it is important to consider the background. Adding depth to a 2D Silhouette Animation is easier than you think. Cutting out hills and castles for the background and then layering them between sheets of paper helps to give the illusion of distance. This technique also lets you animate the background scenery, such as moving clouds, frame by frame between the sheets of paper.

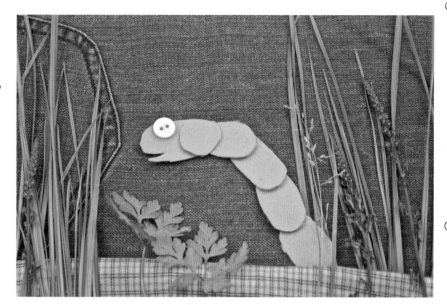

**Top:** Cutout animation made using fabrics and plants.

**Bottom:** Cutout animation made using paper.

# Eimhin McNamara
# *THE ROOSTER, THE CROCODILE, AND THE NIGHT SKY*

In 2007 Eimhin McNamara started work on the award-winning animated short *The Rooster, the Crocodile, and the Night Sky*, which depicts the story of a rooster who goes on a journey to battle a crocodile in order to return the night sky to its rightful place.

Eimhin was first attracted to animation because of the fact that it draws on all the different creative disciplines, from illustration to film, fine art to graphic design, poetry to dance. Choosing Stop Motion as the area of animation he wanted to practice in, Eimhin describes the reason behind this choice:

"Stop Motion has one major point in its favor, which is that anyone with a camera can do Stop Motion. If you look at trends in independent animation and music videos over the last few years, quite a big 'underground' Stop-Motion scene has emerged from this DIY mentality— Fleet Foxes' music videos, for instance, as well as the likes of Mikey Please and Johnny Kelly."

The production of the "Rooster" film took about 13 months, with the first two and a half being spent making storyboards, animatics, and finalizing the character designs. The rooster's character model is made up of 16 different pieces that articulate separately.

The characters were built using a mixture of acrylic paint and inks on different types of poster board. For the nighttime scenes Eimhin made the character in silhouette with black construction paper and colored crepe paper detailing. He wanted the film to have a dreamy, otherwordly feel, and the idea of mixing media came out of this; he also felt it would be a nice way to get the background elements to come alive, different from the characters in the film. On the task of combining the different elements of the film, Eimhin explains:

"Due to integrating so many different techniques it would have been too time-intensive trying to do it all under the camera; in the end we scanned the character elements (over 982 parts for the rooster alone) and compiled the scenes together in Adobe Flash for the animation, and then composited the finished film using Adobe After Effects.

"It would have been much more straightforward to just do it under the camera, rather than faking it on the computer, and I feel the aesthetic would have gelled much better too. Even if we had made this film two years later it would have been much easier to produce it in Stop Motion under the camera, using software like AnimatorHD and using DSLR cameras with live view capabilities."

The animation's story is as unique as the musical score that accompanies it, and the hand-built, digitally manipulated 2D puppets add a unique charm not often found in other fully digital animations.

"The thing I love about Stop Motion is that it is so much faster than any other types of animation. There's a real energy when you're animating something under the camera, an improvisational spark that adds a touch of magic to the process."

**Clockwise from top left:**
- Paint on glass with digitally animated cutout character.
- Rostrum setup for capturing frames of painted sky.
- Close-up of painted sky on glass with backlight.
- The crocodile, made from construction paper, paper, foil, and paint.
- Digitally scanned painted construction paper against background of real leaves and flowers.

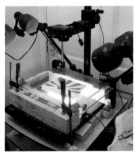

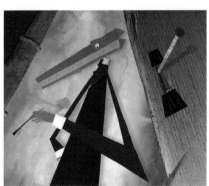

# Javan Ivey
# *MY PAPER MIND*

Having been excited about Stop Motion Animation from an early age and inspired by the likes of Ray Harryhausen and the animated sections of *Sesame Street*, Javan Ivey developed a technique within Stop Motion that has been replicated around the world due to its unique visual look.

> **"**
> Computers did neat light things, but Stop Motion could make things come alive ... Things that weren't supposed to move were made to move ... life for the lifeless. That was the best magic trick.
> **"**

*My Paper Mind* is a short animation that shows objects of childhood memories from Javan's mind and abstracted patterns that appear and then cascade from view as the frames move away from the camera.

This technique involves animating images using traditional 2D drawing techniques onto paper and then cutting out the negative space of the images. The frames are mounted on a custom rig that hangs a sequence of frames in front of the camera. A picture is taken, the sheets on the rig are moved backward one space, and the next frame is put at the front of the rig.

Javan filmed the entire piece on a video camera hooked up to iStopMotion. The rig that holds all the frames was custom built and three poseable desk lamps were used to light the set. The whole process for this animation took about two weeks. Javan spent 24 hours cutting out the frames and eight hours shooting the sequence.

The technique is so visually unique that it has been replicated for use in music videos, commercials, and other short animations. However, understanding the nature of ideas, and how there are always better ones to find, Javan has an interesting viewpoint on how his developed technique has been taken up by others:

> **"**
> I've set that project loose on the world, revisited it once, but I don't have much of a desire to do more with it. I've had too good a time watching other people use the technique and run with it.
> **"**

**Top:**
- Long grass blows in the wind.

**Center (from left to right):**
- Abstracted shapes fall away from the camera.
- The motion arc of the swing can be clearly seen.

**Bottom (from left to right):**
- The custom rig used to hold the hand-cut frames.
- The camera looks "through" the frames.
- What the camera sees.

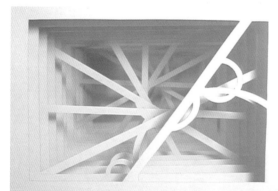

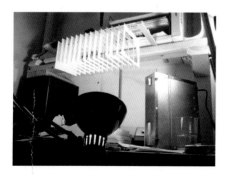

# Camille Baladi and Arnaud Roi
## *DES POP ET DES UP*

Camille and Arnaud are two French pop-up book artists who created an animation made totally from paper. Light and shadows help define the shapes of their characters and sets along with some pixilation photography, which helps to create an enticingly visual animation.

" We wanted to make an animated film with real pop-up effects in it because we are paper engineers (we make pop-up books) and were tired of seeing movies with special-effects pop-up techniques. "

Having designed the set and planned the animation using Adobe After Effects software, Camile and Arnaud photographed an actor (Nicolas Tirard) in front of a white screen, animating him like a puppet. The photographs were cut out and each one was photographed on the set in sequence.

Three spotlights and a camera flash were used to light the set, and a Canon 450D digital SLR camera was used with a Sigma lens to shoot the animation. All the frames were assembled with sound using Adobe Premiere. Marine Le Dantic was hired as an intern to work on the project because of her knowledge of and experience with animation. Marine helped to make the storyboard, oversee the production, and edit the final animation.

From beginning to end the entire animation took three months to make, with Camille, Arnaud, and Marine often working between nine and 19 hours each day.

" If you think about the effort and time spent, we would never do animation! But the pleasure of seeing the pictures come to life is so huge that it makes you want to start again! "

Camile and Arnaud continue to create their pop-up books, but should they experience a quiet period in their everyday work, they might consider another animation project like *Des Pop et des Up*, saying that their experience of making the animation gave them the confidence to try it again.

**Top:**
• A large animation set requires a lot of light.

**Center (clockwise from left):**
• Nicolas Tirard being photographed for his animated sequences.
• The sequence of photographs shot.
• A printed frame of Nicolas cut out.
• Beautiful hand-cut background pieces.
• All the hand-cut frames organized and stored for shooting.
• A final frame from the animation.

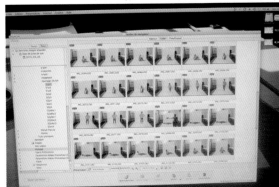

# CREATING A PAPER PUPPET

Regardless of the materials you will need to create your puppet, the main tools always needed for this job are a good pair of sharp scissors, a scalpel, and a cutting mat. Always consider the size of the area you will be animating on and how big your character needs to be in the shot.

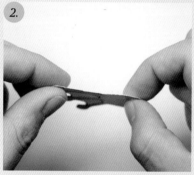

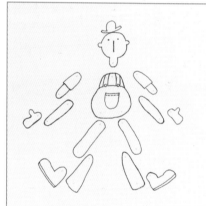

## Sketch it out

Create a sketch of the character first in a "T" pose so that you get an idea of the size of the different parts. The "T" pose will help you to see how long the arms and legs will need to be when creating those parts.

## Articulation

How articulate your puppet will be depends on how many parts it has. For example, do the hands need to move? Don't forget to extend each piece a little longer than your original drawing so that there is some overlap for creating the hinge.

## Making a puppet

1. Color in and add detail to your puppet before cutting out the pieces. This allows you to go over the lines and hold the poster board as you color.

2. Once you have cut out the individual pieces, take some time to straighten the poster board again, as it will bend as you cut it with scissors.

# HINGING A 2D PAPER PUPPET

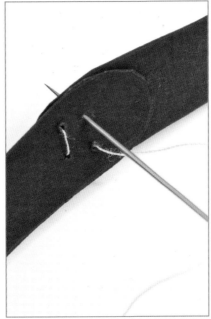

### Split-pin method

Split pins are a simple way to join the pieces of your puppet together. The only downside to this method is that split pins are quite large and can damage delicate joints. They are best used for larger puppets and silhouette puppets because their size means the puppet cannot lie completely flat on the surface on which you are animating.

### Sewing method

You can sew the joints of a paper puppet together using a needle and thread. This is a little more difficult to achieve than using split pins, but the joints of the puppet will move more freely and last longer. When piercing though stiff poster board, place some reusable adhesive or clay underneath the poster board to make it easier.

### Reusable adhesive method

I find this to be the best method for joining the parts of a paper puppet together. The joints remain stiff and in place but are also easy to move and rotate. The join is also invisible when animating, and if you need to replace pieces, you can do so quickly, without interrupting your animation.

# SETTING UP A SILHOUETTE ANIMATION

If you want to shoot a Silhouette Animation, you will need to use a light box. Light boxes are available to buy but can be quite expensive, especially in large sizes. It is inexpensive to make your own light box, and you will have control over its size.

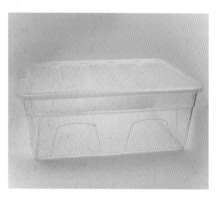

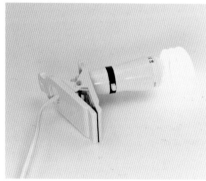

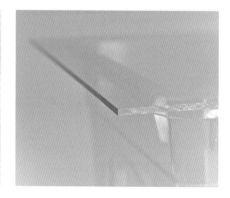

### Choosing a box

Any sturdy box can be used for the base of a light box; just check that the opening is large enough because this will be the area on which you will animate. Try to avoid using cardboard boxes because they will move more easily than a wooden or plastic box.

### Light source

Avoid using hot lights such as incandescent or tungsten bulbs because these will create a fire hazard when they are placed inside a closed box. Instead, use LED lights or an eco bulb to provide the light source in your light box.

### Glass or Perspex?

Using glass for a light-box lid is inexpensive but fragile and dangerous; it could crack or shatter. Perspex is more durable and available to buy online. Try to use Perspex that is ⅛–⅕ in. (3–5 mm) thick so it doesn't flex when animating.

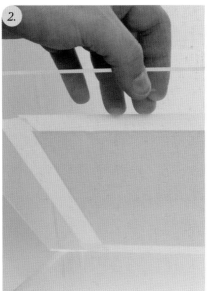

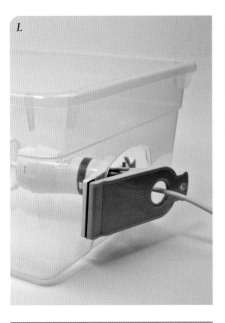

## Building a light box

A plastic storage box with the lid removed makes an ideal light box.

### 1.

Cut some holes around the sides of the box to allow in air to keep the light bulb cool. Cut a larger hole for the light to slide into the box. I am using the clip light as mentioned in Chapter 2 *(see page 21)*, which holds itself in place.

### 2.

Frosted or white Perspex can be too opaque for low-powered LED or fluorescent lamps to shine through. It is best to use transparent Perspex because you can control the opacity by using sheets of paper. Simply tape some paper or tracing paper to the underside of the Perspex to make the lid more opaque.

### 3.

Secure the Perspex lid to the top of your light box with some duct tape. If the switch for your light is inside the box, you could put duct tape along one side of the Perspex to create a hinge to allow you to open and close the box.

### 4.

If there is excess Perspex going over the edge of the light box, you can use it for storing puppet pieces while you animate.

# SHOOTING A SILHOUETTE ANIMATION

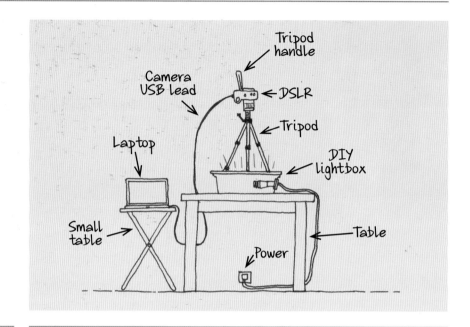

Tripod handle

Camera USB lead

DSLR

Tripod

Laptop

DIY lightbox

Small table

Table

Power

We are going to look at creating a basic shot for an animation from the story *Aladdin*. Fairy tales and children's stories lend themselves well to Silhouette Animation; the stories are so well known that the visual style can be simplistic and the story will still come across.

## The setup

Any down-shooting setup can be used for this type of animation; however, I am going to use a DSLR attached to a laptop running Dragon Frame so that I have good control over the exposure on my camera. This is important when animating on a light box, but most webcams and digital cameras will also let you adjust the camera exposure setting.

With the tripod and camera in place on top of a table, I am going to attach the light box firmly to the tabletop with some duct tape. Be careful when using duct tape on furniture; its glue is quite strong and may pull off varnish or paint when removed.

I have the puppets and scenery already prepared and I am also going to use some play sand when animating the smoke for the genie coming out of the lamp.

**1.**
Place any scenery behind a sheet of paper to create the background and give the appearance of depth.

**2.**
Use a small piece of reusable adhesive on the characters' bodies to keep them in place when animating. If your character needs to move, you can still lift it up to move it around.

**3.**
Use Dragon Frame to adjust the aperture and focus of the camera so that both the puppet and the light from the light box can be clearly seen.

**4.**
If you need to animate your character turning around in the shot, simply take a frame with it facing one way and then flip the puppet over the other way and continue animating.

**5.**
When animating with sand on a light box, use a brush to move the sand around. For safety, cover your keyboard with plastic wrap, avoiding ventilation holes, and cover the computer mouse in a small bag.

**6.**
When articulating the puppet, hold pieces you don't want to move, such as the feet, and move the rest of the body. When animating, you will find that some of the puppet parts will naturally move as you move other pieces.

# *LEGO animation*
# OVERVIEW

LEGO animations, also known as Brick Films, are a great way to experiment with animation because they utilize items readily available in toy stores. Because of the tight-fitting components and ability to stay still, LEGO lends itself well to the requirements of an animation medium.

## Brick types

Many different-shaped LEGO bricks are available, all in varying colors and sizes. The bigger the collection, the more possibilities you have for items to include in your animation.

- New bricks can be bought in buckets or as part of a LEGO set. Buying sets is a great way to pick and choose the pieces you want.
- Second-hand bricks can be found in yard sales, thrift stores, and online. Random assortments of second-hand LEGO bricks can be purchased by the pound online.

## Mini-figures

Having many fashion combinations, the LEGO mini-figures come with an abundance of props and accessories. Look out for different faces; these can be handy when creating different characters for animations:

## Baseplates

Baseplates come in many different sizes; the most common and practical are the 32x32 baseplates (the numbers relate to the number of connectors along each side). They come in various colors and it's good to have a selection available for animation.

## LEGO setups

LEGO sets are like miniature dioramas, each depicting a theme or scenario, so the most common setup when animating with LEGO is to build a traditional set with floor and back or sides. This also makes it easier to animate and work with because you can attach the props and pieces to baseplates to hold them securely when taking frames. This type of setup doesn't require a lot of space and in most instances can be assembled on a small table or desk. Some Brick Films use LEGO baseplates as the backgrounds for their sets, whereas others use colored paper or drawings.

Searching for LEGO animations on the Internet can reveal quite a few interesting setups. One of my favorites is the video *8-Bit Trip* by Rymdreglage on YouTube. In the video, LEGO pieces are used to recreate computer games such as Pac-Man and characters from Mario Brothers. The video also utilizes the LEGO bricks to create abstract patterns and sculptures, all moving in time to music.

# Michael Hickox
# *LEGO MINI GOLF*

Michael Hickox has created quite a large online following for his LEGO movies that utilize mini-figures, props, and buildings to create unique short animations that are an inspiration to other animators.

Michael started experimenting with animation by using a digital camera to animate people and action figures before switching to LEGO. His decision was based on the popularity of his LEGO animations online compared to the other creations.

**"**

I produced videos to satisfy my fans, so I began creating LEGO videos exclusively. I felt that I was creating my videos for others to watch, so I had no objection to following what they wanted.

**"**

This ability to monitor the popularity of his animations has inspired Michael to create more than 25 shorts.

"I have generated a fairly large following on YouTube, with about 40,000 subscribers, and over 80 million views throughout my five years online."

Having come up with the idea for *LEGO Mini Golf*, Michael searched the Internet to see if anything similar existed. Part of his ethos is to try to create something unique so that it attracts viewers.

"A good video must balance animation, a storyline, computer editing, and creating something that will stand out. When I'm thinking of a video, I'll ask myself, 'Why would someone want to watch this?' "

Michael shot *LEGO Mini Golf* in his basement, using a fluorescent lamp and a Panasonic digital camera, and edited it in Sony Vegas. Michael takes each frame while standing behind a poster board wall to minimize interference with the lighting.

This animation took around 10 hours of set building, 30 hours of animating, and ten hours of editing. Animating takes a long time because the mini-figures can only move about 1/8 in. (3 mm) each time. When animating walks, Michael uses five different positions for each step.

Most people would understandably think that LEGO animations are for a young audience; however, their appeal exists across all ages. As Michael explains:

"Although my videos are intended for a young audience, YouTube viewer data supports the fact that people of all ages can appreciate LEGO Stop Motion."

**Top (from left to right):**
- Animating the golf ball requires digital masking to remove the support.
- Mini Golf is anything but relaxing.

**Center (from left to right):**
- The ball strike should imitate natural motion.
- Change the figure heads for different emotions.

**Bottom (from left to right):**
- The windmill had to be moved every frame throughout this shot.
- Small, clear pieces are used to simulate flowing water.

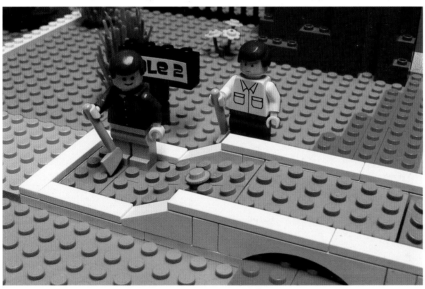

# Kevin Horowitz and Jacob Schwartz
# *WRITER'S BLOCK*

Being stuck for an idea is something we have all encountered, but the solution is usually staring us in the face. This is the premise for *Writer's Block*, an animation by Kevin Horowitz and Jacob Schwartz that uses dramatic lighting and surreal sets to tell the story.

After producing a series of comedy films, Kevin and Jacob wanted to create something more dramatic and artistic, so they decided to create *Writer's Block*. Also, motivated by their high school's annual film festival, they went on to win local and international film festivals with their movie.

Using LEGO pieces they already owned, the pair started building sets and characters needed to shoot each of the scenes. Kevin spoke about the experience:

"One of the main advantages of using LEGO is that almost every kid has some, and so anyone can start animating without needing to buy a ton of supplies."

Working individually, they captured the frames using Monkey Jam and a web camera that allowed them to film both close-up and wide shots of the mini-figures.

> " Web cameras are actually really great for animation compared to traditional video cameras because they allow for extreme close-up focus and are able to physically be put in tight areas in the set. "

To keep flickering to a minimum, the sets were built on a desk placed away from any windows so the natural light didn't affect the work, and everything was held down with earthquake putty, which is strong, cheap, removable, and reusable.

The movie took five months to produce, with both Jacob and Kevin animating for an hour a day. As shots were completed, Kevin started to edit the movie together using Adobe After Effects. Some CGI concept shots and shots of the movie were created using a free trial of Autodesk's 3D Studio Max.

The time and effort taken to create *Writer's Block* provided a useful experience. Having also experimented with 2D Drawn Animation, Kevin hopes to create more animations.

"Even if I have to restart or completely scrap a project, the experience just increases the quality of the next project."

**Top:**
- A 3D CGI concept shot for the movie.

**Center (from left to right):**
- The same shot as realized in the final movie.
- Dramatic lighting used to enhance the atmosphere of the movie.
- The shape-shifting blob.

**Bottom (from left to right):**
- Cellophane used for water.
- After Effects was used to create the distortion visuals.

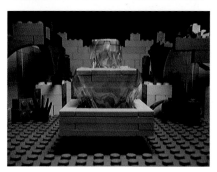

# Tomas Redigh
## *8-BIT TRIP*

Since its release in 2009, *8-Bit Trip* has pulled in millions of views on YouTube. Its success is down to retro nostalgia mixed with very obvious hard work in animating thousands of LEGO pieces in various patterns and familiar 8-bit graphics. A massive undertaking, this animation has inspired many others to try LEGO Animation for themselves.

Through the success of *8-Bit Trip*, Tomas Redigh is often asked why he didn't just create the animation using CG software:

"The answer is because it would take a lot more time. Moving all those bricks in the computer and getting the same result with all the little defects which come with real LEGO bricks would be nearly impossible."

*8-Bit Trip* was inspired solely by old video games, LEGO, and the retro vibe that it uses throughout the animation. It was shot on a Canon D90 DSLR using both a variable zoom lens and a macro lens for close-up shots.

Tomas used high-frequency fluorescent lights, which are more expensive than standard ceiling fluorescent lights, but help to reduce the flicker that ordinary ceiling lights can sometimes produce when used for animation.

Camera movements were created by rotating the set Tomas had built, rather than rotating the camera. He placed a measuring tape along the edge of the table so he could measure each turn precisely as he animated.

Tomas used 25 boxes of LEGO over an eight-month period to create *8-Bit Trip*, and the job took him around fifteen hundred hours. While this seems like an extraordinary amount of time to create a three-minute and fifty-second animation, the effort is clearly evident and very much worth it.

> " For me, animation has a high reward factor, but I have been lucky with my *8-Bit Trip*. I do ordinary movies that also often take a lot of time, but I do not get as much reward from those. "

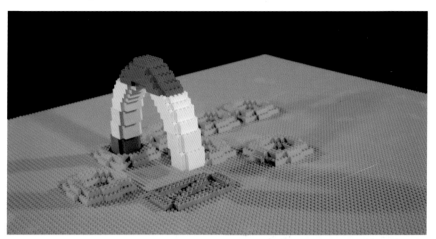

**Top:**
• The snake from the beginning of the video.

**Center (clockwise from left):**
• Kung Fu game video game art.
• A video-game loading screen built on nine large LEGO baseplates.
• The LEGO snake replacement "frames" built from LEGO.

**Bottom (from left to right):**
• The actors get transformed into the 8-bit world.
• The custom-made rotating table used for camera movements.

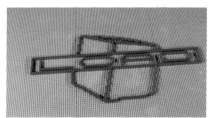

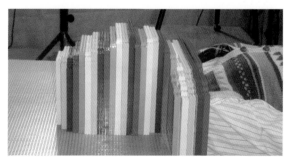

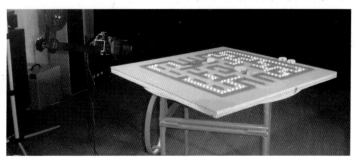

# BUILDING A BASIC LEGO SET

With any animation, it is good to have a clear idea of what you want to happen before you start out. Your idea can influence the type of set that you need to construct and what it will look like, and there are a few things that need to be considered when building a LEGO set.

### Working with mini-figures

- Mini-figures are only 1½ in. (4 cm) tall, so consider this scale when planning your LEGO set.
- Remember to place small items in the background to help create the illusion of distance.
- If you are planning to change the expression of a mini-figure, check that you have all the necessary heads.

### Themes

LEGO sets come in different themes, such as Space, Town, Castle, and Pirates. You can base an idea on one of these or try to create your own theme by using different bricks and mini-figures.

### Props

Many LEGO props exist to use in your animation, ranging from axes to ATMs. Remember to consider background props, such as trees, which help to add depth to your set and to fill in the background of your shot.

## Keep it simple

Building an elaborate set piece is good fun, but when it is for use in an animation it is best to keep it simple and small scale. You also need to consider how much will be visible by the camera when it is placed on set. Most webcams have a narrow viewing angle, so not everything you build will be seen in the shot.

## Use the baseplates

LEGO baseplates make a perfect "ground" for all your figures, buildings, and props to attach to when animating. You also need to consider the background because this will take up a large area of the picture. If your animation idea is based outdoors, then a good way to create sky for a set is to use a large sheet of thick blue paper.

Tape one edge of the paper to the table or desk and then curve the rest onto the wall behind. This creates what is called an infinity curve *(see page 39)*. If the curve is big enough, you won't see any crease when looking through the camera, giving the appearance of a sky in the background of your shot.

# SETTING UP A LEGO ANIMATION

Once everything is ready to shoot your LEGO animation, make sure you set up with plenty of room to work. When using software to capture frames, it is best to keep the computer away from the set on a separate table or chair. This is because you will need room to move around the LEGO set and place the webcam for each shot.

### Check all props

Ensure you have everything you will need to be in each shot. Stopping for long periods of time between frames to build something is not a good idea.

### Use a storyboard

You may need to reference the previous or next shot in the scene, and the easiest way to do this is to reference a storyboard for the animation.

### Securing props and lighting

Any props that aren't secured to a baseplate should be held down with reusable adhesive. LEGO pieces are very small and light, so it can accidently move even if you aren't animating it.

Secure the lights to the table you are animating on or to light stands. LEGO pieces are highly reflective, and if a light moves in between your frames, it will show up in the final shot.

## Checking and setting shots

Once the set has been constructed and the lights are in place, connect the camera you are shooting on and make sure that you can get the different shots you want to be in your animation. This is important, as you may find that you need to move your set slightly on the table or that you need to mount your webcam on a tripod for one of the shots. Knowing all of this before starting to animate can save time later on.

When animating with a webcam, it is important that you have control over settings such as white balance, exposure, and focus. Leaving these settings on automatic can ruin a shot as the webcam will try to change these settings continuously, resulting in a flickering shot. Open the webcam settings beside the capture software window so that you can make adjustments as and when you need to at the start of each shot.

This setup is also handy if you want to animate a focus shift during an animation, where the camera changes its focus from one item to another. Simply change the focus by a small amount on each frame as you animate.

# SHOOTING A LEGO ANIMATION

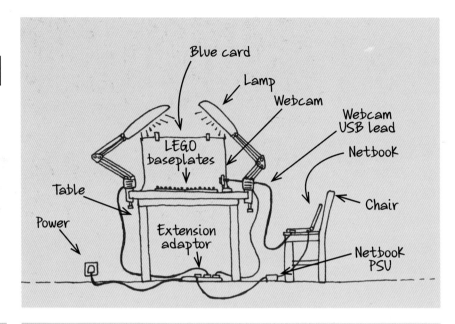

For this tutorial we are going to animate different shots of a scene about a car race between two LEGO mini-figures. This means we will be relying on a storyboard to help keep track of the different shots and which camera angles we will need. I am using two green baseplates to build a basic set using small LEGO trees that, when set in the background, will look like they are farther away than they really are.

## The setup

We will be using a small table to set the LEGO baseplates on and a sheet of blue poster board for the sky. Two fluorescent anglepoise lamps will provide the lighting, while everything will be shot on a Logitech C615 Web camera. The camera will be mostly supported by reusable adhesive as we need to keep the camera quite close to the subject matter throughout shooting, and even a tabletop tripod would be too big and clumsy to manage.

I am going to use Helium Frog to capture the frames along with the webcam software to control the exposure, focus, and white balance. The software will be run from a Netbook connected to the web camera.

### 1.
To minimize movement of the cars as you animate the characters inside, mount them on some bricks, especially during close-ups. This means that you can animate the mini-figures without accidentally moving the vehicle.

### 2.
Mini-figure helmets are detachable, which makes it tricky to animate heads turning. Simply take the helmet off, move the head, and replace the helmet, lining it up with the face.

### 3.
Animating a walk is trickier than you might think. After placing the legs in the correct position, place the heel and toe of each foot in between the baseplate connectors. Use a small piece of reusable adhesive to help hold the figure up while you take the frame.

### 4.
To give the illusion of the car moving along at speed, I animated the trees in the background moving in the opposite direction to the car.

### 5.
For the last shot where the car flips and flies through the air, because of the angle of the shot, I simply held the car with my hand. For each frame, I made sure that my hand wasn't in the shot and that it wasn't casting a shadow on the set.

# *Post-it Note Animation*
# OVERVIEW

Post-it Note Animation is something of a new phenomenon. Replacing pixels with paper squares lends itself well to Stop Motion Animation and has influenced people to try out animation by creating short movies with the square notes. The ability to create amazing pieces of work using an everyday item such as a Post-it note is something anyone can try.

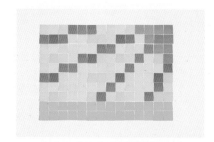

The most popular Post-it note color is yellow; however, there are many other colors in the Post-it note range to choose from to create the images you want. You can also paint and color the Post-it notes to create custom-made color palettes to work with.

The most common shape of a Post-it note is square; however, there are large and small notes to choose from, as well as different shapes. You can also cut out and create your own shapes from Post-it notes, or you can combine different notes together to create new shapes.

### Enhancing the possibilities

Using multiple Post-it notes, you can build up an image of anything you wish. This is the same principle used to create all the imagery seen on a computer screen, where hundreds of pixels make up the image, like a mosaic picture.

You can also draw on the Post-it notes themselves to create more detail or to do some 2D Drawn Animation. Mixing Stop Motion Animation with Drawn Animation can result in some interesting visuals. It is worth experimenting with the technique.

Post-it notes can be stuck to most surfaces, which means you can create a Post-it Note Animation almost anywhere you wish. The most common location to create this type of animation is in a studio or office environment, but it is worth trying out other locations, such as outdoor windows and walls. Animating Post-it notes by themselves on a wall can be fun.

You can also try making the Post-it notes interact—or seem to interact—with people or other objects, sparking new ideas and stories for your animation. For example, try having them form a flower growing out of a vase or become a pair of curtains for a window.

**Top:** Post-it note shark design on an exterior wall.
**Bottom:** Post-it notes used to create a paper flower in a real vase.

# Jeff Chiba Stearns
## *ODE TO A POST-IT NOTE*

"A Post-it note escapes on an incredible journey of self discovery to find its 'father,'" reads the synopsis for this short movie by Jeff Chiba Stearns. Produced for 3M Canada, the award-winning *Ode to a Post-it Note* utilizes Stop Motion Animation to tell the story of one note's journey of discovery.

A studio or office space is the typical location to animate in; however, Jeff shot this animation in chronological order and actually animated in the different locations on his journey, taking out his DSLR and shooting wherever he found himself next.

> I started in my office, and continued from there to the airport ... I actually shot all the airport scenes as I was waiting for my flight.

The shoot was finally located in St. Paul, Minnesota, at the 3M headquarters where Jeff animated the inventor of the Post-it note, Arthur Fry, playing with his own invention. However, animating outside was a challenge in itself.

"When I was animating the Post-it note walking up to the 3M Innovation Center, it was super windy and many times my Post-it note hero would blow away. It was a difficult shoot."

The entire animation was shot with a Canon 40D DSLR and a wide-angle fish-eye lens, and then edited together using Final Cut Pro. Jeff took 50,000 photographs during the animation and edited them down to 3,500 for the final piece. It took three months to complete the movie, with one month spent on pre-production, another spent animating, and a final month of postproduction.

Jeff then worked on another animation project entitled *Yellow Sticky Notes: Canadian Anijam*, incorporating the work of 15 award-winning animators from across Canada. It's clear that Jeff will continue to do Stop Motion Animation for the foreseeable future.

> People can appreciate the time and technique of creating Stop Motion Animation even though they can't figure it out. Animation is one of the most misunderstood art forms to ever exist. It really is an amazing magic trick, bringing inanimate objects to life.

**Top (from left to right):**
- Jeff really likes Post-it notes, and uses a lot of them.
- Jeff with some of the notes used for another animation.

**Center (from left to right):**
- The little Post-it note starts its search.
- The Post-it note discovers its own father.

**Bottom (from left to right):**
- Outside the 3M Innovation Center.
- Jeff with the inventor of the Post-it note, Arthur Fry.

# James Sturton
## *POST-IT STOP-MOTION THANK YOU*

Saying thank you can be done in so many different ways, but taking the time to say thank you through the medium of animation makes it seem all the more genuine. James Sturton creates a YouTube series called *Video Games with James*, and to thank his many subscribers, he decided to make a special video. As James says:

> " I have used so many techniques in the past throughout my videos, including green screening, and I felt it was time to try something different. Stop Motion Animation looked really appealing and fun, and I wanted to give it a go. "

Starting out with 940 Post-it notes, James went about building a frame within which an 8-bit character runs along while different subscribers' names appear on the board above. Some Post-it notes eventually lost their stickiness, so James ended up using an additional 60 Post-its as replacements.

The animation was created on a wall in James's lounge and was lit using an Interfit lighting set to keep the light while he was animating. The frames were shot using a Canon 450D DSLR with a wide-angle lens to help fit the entire wall into the frame. The camera was triggered using the Canon Remote Capture software, and the image sequence was then edited using Corel VideoStudio Pro X4.

The entire thank-you video took around three weeks to complete. This time was split into two weeks of animating (working about nine hours a day, every day) and one full week of editing. In total, James spent 200 hours on the project, but this has simply encouraged him to do more:

> " I have actually planned on making a Stop-Motion video based on the video game Tetris, and even Donkey Kong, inspired by the running man I created that gave a sort of 8-bit NES animation effect. "

This project is a unique way to utilize Stop Motion Animation. Because of the amount of time and effort required, animation is usually reserved for telling a story or idea, but to use it for saying thank you is very special indeed.

**Top (from left to right):**
- James in front of the wall he will use for the animation.
- Each Post-it note is individually put in place.

**Center (from left to right):**
- Fully constructed, the animation frame is almost as tall as James.
- The word "record" comes down from the top.

**Bottom (from left to right):**
- Changing Post-it notes for the next frame.
- The character crosses the finish line with James's YouTube channel name.

# Aaron Kaminar
# *IF YOU EVER NEED SOMEONE—*
# *THE FAMILY BONES*

There are few doubts in the viewer's mind how a Post-it Note Animation has been made when they are watching it. We can see the vast amount of work and effort that have been put into the piece, as we are familiar with the Post-it note as an everyday object. This music video by Aaron Kaminar from Image Flux is a good representation of creativity mixed with hard work and dedication.

Having worked in postproduction as an animator, compositor, and visual effects supervisor for many years, Aaron was looking for a new project to direct when he met Justin Webb from The Family Bones.

> " I met Justin Webb of The Family Bones while looking for a project to direct. We hit it off creatively and came up with the concept for the video together. "

Planning was key to this project, so Aaron used an animatic to time out the storyboard in time with the music to make sure he knew how long each shot needed to last. The animatic was also used when it came to animating the Post-it notes on the wall of a warehouse.

This process involved the projector being switched on, the Post-it notes being aligned with the projected image, the projector being turned off, the studio lights being turned on, and the frame being taken on the Canon 7D DSLR.

Having access to a warehouse for the project meant the area to animate on was huge, and the largest frame consisted of 2,304 Post-it notes. Final Cut Pro was used to edit the piece together, and the entire project, from conception to completion, took about four and a half months, with most working days lasting about ten hours.

**Top:**
- The amazing singing sequence.

**Center:**
- Pixilation on top of Stop Motion Animation.

**Bottom (from left to right):**
- A wide-angle shot of the warehouse wall.
- All the required colors sorted to help speed up production.
- Neatly placed notes.

# CHOOSING YOUR POST-IT NOTES AND LOCATION

Because Post-it notes come in so many different shapes, sizes, and colors, it is good to work out roughly how many you will need for your animation. Creating a basic picture to start off with may only require a certain number of notes, but as you change each frame, you may find that you require more notes of a particular color.

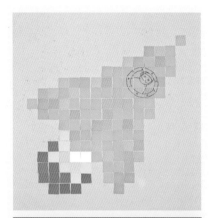

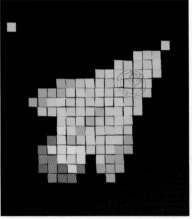

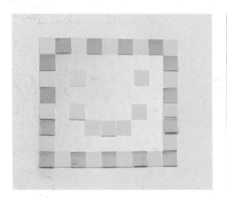

Using large Post-it notes will help to reduce the amount of notes you have to move each frame. It can also help save on the amount you will need. Even a simple image, such as the one above, takes 37 Post-it notes.

As well as requiring a lot of notes for the animation, you may need to replace used notes because the sticky strip can become worn and not stick anymore. The number of times you can reuse a Post-it note will depend on the surface to which you are sticking it.

Try to pick a surface on which to animate your notes that contrasts with the general color of the notes that you have chosen to animate with. A white wall is good, but for some animations, a black wall will help them to stand out better. Try to avoid patterned or textured walls because these can distract from your animation frames.

## Shooting outside

Creating a Post-it Note Animation on an outside surface can result in an interesting visual look, with the natural light changing each frame. Be aware that you will be at the mercy of the elements; if you start late in the day, you may not finish the animation before the sun goes down. As Post-it notes stick only along one side, you will also be at the mercy of the wind, which can easily blow part of your work away.

## Room to move

Give yourself lots of room to animate and don't forget to consider the height of the image you are creating. Make sure you have a stepladder if you need it and try not to animate on a wall that has furniture or desks in front of it. For example, while an animation happening on the wall of a living room may look good, remember that you will have to climb over the furniture to change the position of the notes for every frame.

# PLANNING A POST-IT NOTE ANIMATION

Sometimes it is fun to just dive right in and start animating a Post-it Note Animation, but it is a good idea to plan out your idea first. Otherwise, you may run out of notes or not have the right quantity of notes in specific colors that you need to complete the animation. You can plan in a number of different ways.

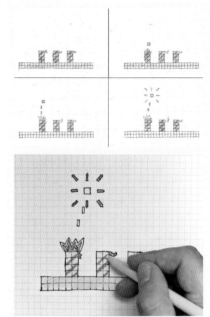

### The uses of graph paper

Having all the frames prepared on graph paper first can help if you are shooting outdoors or are working in a group of people. Each drawn frame on the graph paper can act like a blueprint for each frame of the Post-it Note Animation. Filling in the squares on graph paper with the colors of the notes you are going to use will also help you to plan the quantities of colored notes you need.

### Sketching

Doing a rough sketch of the images you want to create in the animation can help you to visualize everything first. If your idea involves lots of things happening, you may want to sketch out many pictures, much like a storyboard.

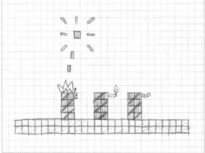

### Graph paper

Using graph paper is a great way to work out exactly how many Post-it notes you will need for your animation. You could also draw each frame of the animation on the graph paper so that you have everything perfectly planned before you start animating.

Using a down-shooting setup with a webcam or your cell phone, you could then take a sequence of images of your graph-paper drawings to check that everything moves as you want. Correcting any mistakes at this stage will be easier than trying to figure out the corrections when you are halfway through an animation. Numbering each sheet allows you to reference each frame of the animation as you create it to make sure it is right before taking the frame.

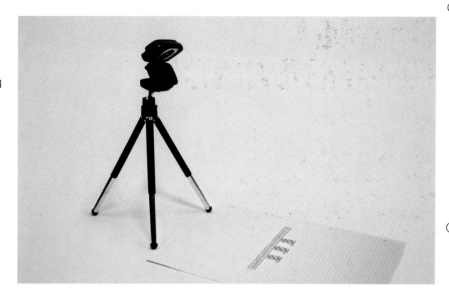

## Poster board panels

If you are working with large images, it can be tedious to change all the Post-it notes for every frame of the animation. Depending on the size of your image, you could create panels of Post-it notes on squares of poster board. Then, instead of moving 81 notes, you could just move nine panels of nine Post-it notes; this takes less time, and the notes will last longer. This can work well when animating characters whose body shape does not change during the animation.

# SHOOTING A POST-IT NOTE ANIMATION

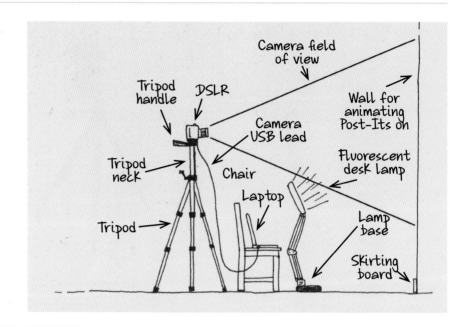

For this tutorial I am going to create a basic Post-it Note Animation indoors on a blank white wall. The animation will consist of a series of fireworks going off, created with Post-it notes, and I am also going to include myself in the animation, lighting the fuses of the fireworks.

## The setup

Because I am going to animate my hand lighting the fuses of the fireworks, I am going to use Dragon Frame connected to a DSLR, which will let me use the high-quality onion-skinning feature in the software to make sure my hand is in the right place for each frame.

I am going to use a fluorescent desk lamp to illuminate the shot, although any type of lighting can be used for this setup. If using a desk lamp, you may want to consider using tracing paper or diffuser paper to help diffuse the light *(see Chapter 2, page 22)*.

I will be using a variety of Post-it notes to create the animation, which I have planned out on graph paper to make sure that I have everything I need before I start to animate. To create the flame of the lighter and the sparks coming out of the fireworks, I am going to cut some Post-it notes into the different shapes required.

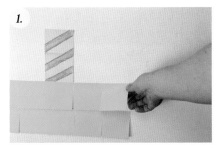

**1.**

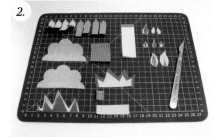

**2.**

**3.**

**4.**

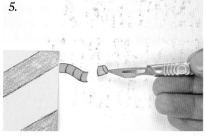

**5.**

**6.**

**1.**
Once you clear a space on which to animate, put any permanent, non-moving Post-it notes on the wall. This will help you position the camera and lights correctly before animating.

**2.**
Put all the small Post-it notes and pieces on a cutting mat to make it easy to access them as you animate. Use a scalpel to help position the smaller pieces, such as the flames, on the wall.

**3.**
The flame on the lighter is animated by swapping two different cutout flames every frame. This gives the appearance of flickering when played back.

**4.**
To make picking small pieces off the wall easier, use the tip of a scalpel blade to help lift them off.

**5.**
To animate the fuse on the firework getting shorter, a piece of the fuse was cut off with a scalpel for every frame.

**6.**
Dragon Frame's onion-skin feature really helps when animating yourself. Holding still between frames is tricky, but the results are usually worthwhile.

# *Smartphone Animation*
# OVERVIEW

Shooting Stop Motion Animation on a phone allows you great freedom to shoot where and when you like because your phone is generally with you wherever you go. The versatility of the apps that are available for phones means you can even experiment with pixilation animation and time-lapse photography.

### iPhone

The Apple iPhone was late to the game in offering decent camera quality. However, as each new generation of the iPhone is released, the camera becomes more and more advanced. This advantage, coupled with the iPhones App store, makes it extremely versatile for using as a digital camera.

### Fingerlab iMotion HD

A popular Stop-Motion iPhone app is Fingerlab iMotion HD, which lets you shoot images at 720p resolution and has an inbuilt onion-skin function that aids Stop-Motion photography greatly. The basic software is free but the full version (which allows you to export the animations you make) is quite inexpensive.

Regardless of which smartphone or app you use to shoot Stop Motion Animation, there are a slew of peripherals and attachments available for most current phones. While the lens in a smartphone is generally fixed, so it can't zoom in or out, you can buy attachment lenses that will add this functionality. If you decide to invest in a lens attachment for your smartphone, make sure you purchase a good-quality lens; cheap models can result in worse image quality than the basic camera on the phone.

## Smartphone tripod mount

The only function missing from any smartphone is the ability to mount it to a tripod so that it can be kept steady while shooting frames. You can buy tripod holders for smartphones, but they can be expensive and become out of date if you upgrade your handset each year.

Building a smartphone holder is an option and requires little more than some wood, a tripod nut, and some wood glue. Cut the wood to the length of your phone and drill a hole for the tripod nut. Then glue the pieces together with wood glue, tighten in a clamp, and leave the holder to set overnight.

If you want extra security, you can use rubber bands to attach the phone to the tripod mount. This enables you to put the smartphone on a tripod and use it for regular horizontal shooting or down shooting.

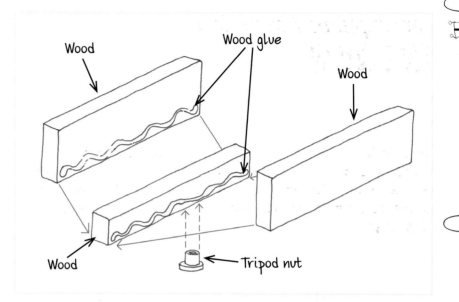

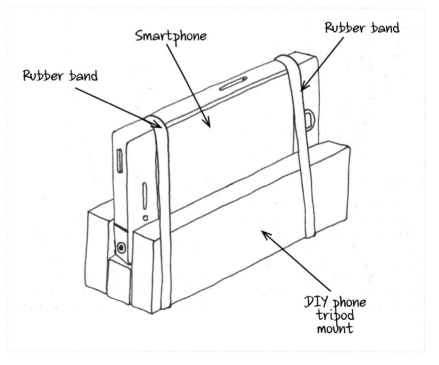

**Top:** How to build your own tripod mount.

**Bottom:** Attaching a smartphone to the tripod mount.

# Will Studd and Ed Patterson
## *DOT*

What can you do when you've done it all within Stop Motion Animation? When they were asked to create a short movie using only the Nokia N8 smartphone, Will Studd and Ed Patterson from the studio Sumo Science decided to investigate miniature objects and create the world's smallest animation using a cell phone.

Working in conjunction with Aardman Animations, Will and Ed have produced a record-breaking piece of work owing to the scale of the puppets used. Their influences while growing up, though varied, are illustrated in this fantastic feat of animation ingenuity.

"We were both inspired by different things growing up. I was really into special-effects animation and title sequences, more graphic stuff."—Will

"From Morph running and jumping around to Ray Harryhausen's skeletons fighting real actors, they all have a presence that drawn animation can't capture."—Ed

Taking twelve hours each day for three weeks to animate, *Dot* came about as an open brief from Nokia requesting a short movie using their Nokia N8 and a microscope. After much discussion about what they should animate, Will and Ed came up with the final idea, inspired by *Alice in Wonderland*.

> "
> We wanted the movie to have a slightly Edwardian vibe, so we chose our props carefully. Things like vintage wallpaper, clockwork cogs, fabrics, crystals, and pins all helped to translate this theme onto the screen.
> "

Working with hand-painted 3D printed replacement characters, everything was set up on a custom-made table top that could move left and right to scroll the background as Dot runs along.

The camera, with its attached "Cell Scope," was housed above the table in a bespoke caddie. Although this setup sounds elaborate, it adheres to a basic setup for Stop Motion Animation.

> "
> Animation, in its simplest form, requires just a camera, a subject, and some lighting. This allows you to focus on the things that are important to you.
> "

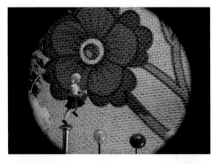

**Top:**
- One of the Dot replacement poses.

**Center (from left to right):**
- Real money used for the background in this shot.
- Pinheads give good reference for the scale of the character.
- Actual ladybug used for scenery.

**Bottom (from left to right):**
- Dot fends off the destruction.
- Knitting the world's smallest duvet.

# Will Studd and Ed Patterson
## *GULP*

When an opportunity comes along to create animation outdoors, things start to look up. Except it's the camera capturing the frames that is up—100 ft (30.4 m) to be precise—above a section of beach as Will and Ed from Sumo Science create the world's largest animation using a Nokia smartphone.

After making *Dot (see pages 138–139)*, Will Studd and Ed Patterson were asked again to make a movie using the Nokia N8 Smartphone, but this time they had to create the *biggest* animation ever made. Nokia wanted sand to be used as the medium, so Will and Ed started to think about how the final piece might look. They explained how they worked:

"We felt the narrative had the feeling of an old English folk story, and we were drawn to the fantastic artwork of Angie Lewin and also the opening sequence to *Watership Down* when we came to the design stage."

Three Nokia N8 smartphones were used to capture frames (which include a full-size fishing boat), mounted on a crane 100 ft (30.4 m) above the beach, where the frames were drawn into the sand. The phones were triggered to capture the frames via Bluetooth using wireless keyboards.

Animating outside means you are at the mercy of how the light changes, although this is something that played to Will and Ed's advantage.

> In the end the weather was fantastic, and the moving shadows (as the day progresses) are one of our favorite aspects; they add an authenticity that we never really envisaged before the shoot.

Working so close to the water is perilous though because a changing tide can quickly come in and ruin a shot:

> At one point we had the team quickly dig a trench around the 'canvas' that bought us an extra frame. Luckily, it was the last frame in the shot and we got it. It was pretty close.

Not long ago, producing a quality animation with a cell phone would have been extremely limiting. With technology and an understanding of Stop Motion Animation, *Gulp* is proof that, when creating animation on a smartphone, imagination is the only limit.

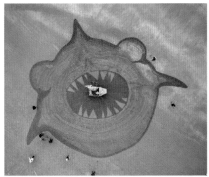

**Top (from left to right):**
- The crane holding the cameras 100 ft (30.4 m) above the beach.
- View from the crane as animators change the frame.

**Center (from left to right):**
- Workers animate the sand.
- The fisherman at sea.

**Bottom (from left to right):**
- Animators come in to animate the "scenery."
- Animators working at night on the mine shot.

# Kontramax
# *THE MIRROR CUBE*

Using Fingerlab iMotion HD App for the iPhone and a self-made camera mount, Kontramax set out animating one of his many puzzles, moving in time to a personally created music track.

Kontramax had originally experimented with computer animation when he was younger, using basic graphic text symbols and then moving on to more sophisticated vector-based animation before practicing in 3D Studio Max.

It wasn't all about computer-based animation though; Kontramax had also experimented with 2D-drawn flip books and was inspired by traditional 2D animations that he had watched while growing up.

> **"**
> The simplified method of presenting the visual information in Stop Motion Animation leaves more space for creative realization of the original ideas. And it looks cool—more 'handmade' than photorealistic computer renders.
> **"**

To create *The Mirror Cube* Kontramax built a stand for his iPhone out of some square metal tubing, foam pads, and some rubber bands so that it could be fixed to a tripod. The animation was shot on a windowsill using natural light and a paper lamp as a second light source to bring some warmth to the images.

The images were shot using the Fingerlab iMotion HD app, allowing the camera to be controlled by sound to take a frame (which is handy when both your hands are also in the animation). It took four days to complete and upload; two days were needed for animating and creating the music, and another two days required for postproduction.

"This very short period for Stop Motion Animation was possible because I used little animation sequences as loops for my project instead of using pure frame-by-frame animation throughout the entire movie."

Kontramax used Sony Vegas 9 to edit the animated sequences together with the sound, which was created using a MIDI keyboard and a desktop computer fan to create the sound when the cube rotates. With more puzzles on his shelf, Kontramax hopes to create some new animations in the future using a DSLR.

**Top (from left to right):**
- The pieces Kontramax used to mount the iPhone to the tripod.
- iPhone set up with the Fingerlab iMotion HD app.

**Center (from left to right):**
- Pencil marks to make sure the cube was always in the same place.
- The mirror cube changes shape to the beat.

**Bottom (from left to right):**
- A grid overlay is used to keep everything in the right place.
- Different stages of the cube animation loops.

# SETTING UP A SMARTPHONE ANIMATION

Because smartphones are self-contained units, and the applications that run on them work directly with the hardware in the phone, you are dependent on the quality of the camera and how well the animation software controls it. There are some setup techniques that will aid you when you are animating with a smartphone.

## The setup

Animating people outside is a great way to achieve fascinating results. When doing so, apply basic photography principles and shoot away from the sun so that the subject matter can be clearly seen. Support the phone as best you can to minimize its movement as you take the frames by propping it on a wall or resting your elbows on a fence.

When shooting animation indoors, it is best to try to mount the phone on a tripod to make sure that it is kept steady. Any type of tripod will do and it is worth experimenting with small tabletop tripods; their size allows you to place the phone almost anywhere to animate from interesting camera angles.

*1.*

*2.*

*3.*

## Down shooting

The smartphone screen can be difficult to see when the camera is at an awkward angle, for example if it is on top of a tripod on a table and you are using the phone to down shoot. You can modify the DIY phone holder shown on page 137 to support an angled mirror, which allows you to see everything on the screen at eye level.

**1.**
Cut two pieces of poster board with 45-degree angles and use double-sided foam pads to stick them on either end of the phone mount.

**2.**
Place a mirror along the two pieces of poster board. This will hold the mirror at a 45-degree angle to the screen.

**3.**
When the phone is pointing downward you can clearly see the screen as you animate.

## Lighting

Smartphone camera lenses are small, and so it is always best to try and supply as much light as possible. Use a strong, white light source if you are using a smartphone to animate indoors. Don't mix different light sources because the camera and camera app will try to white balance automatically as you shoot, which can change the colors of each frame. This means the colors will flicker and change when you play back the animation.

# DOWN SHOOTING WITH AN iPHONE

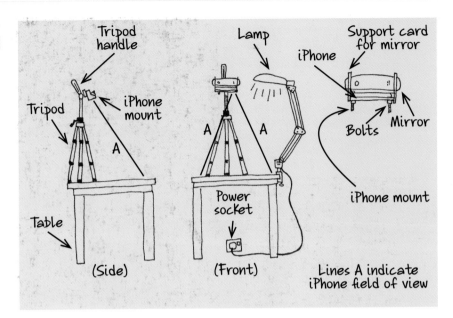

Tripod handle

iPhone mount

Tripod

Table

A

(Side)

Lamp

iPhone

A — A

Power socket

(Front)

Support card for mirror

iPhone

Bolts — Mirror

iPhone mount

Lines A indicate iPhone field of view

Due to the versatility of using a smartphone for creating Stop Motion Animation, it is important to decide and plan which sort of shot you want to create. Do you want to create something elaborate? Something simple? Do you want to use crafted characters or use something bought or found?

You will also need to consider if you want to create a down-shooting setup or shoot horizontally into an animation set. Of course, you can also take your smartphone outside and create some time-lapse-based animation or pixilation. In this tutorial we are going to create a basic "down-shooting" setup to animate some everyday objects as if they are fish swimming in a deep sea.

## Tools and materials

For this tutorial I used an iPhone, along with a piece of software called iMotion HD by Fingerlab, which is available in the app store for free.

You are going to need a smartphone camera mount, which will be used to keep the iPhone secure on a tripod *(see page 137 for how to build a DIY phone tripod mount)*. The objects that I have chosen to animate in this example are:

- Some bulldog clips of various sizes
- Colored paper clips linked together
- Multicolored party balloons (deflated)
- A corkscrew
- Colored Popsicle sticks and craft butterflies for background decoration

## The setup

For this animation setup, I recommend using a basic table with plenty of room around it for maneuvering. Clamp a basic desk light to the side of the table and lay down a sheet of black poster board or paper to use as the background. Put the tripod with the iPhone camera mount attached on top of the table so that it is pointed at the black paper. You will notice that I have placed the tripod slightly off center in order to accommodate the fact that the lens of the camera on the iPhone is set to one side.

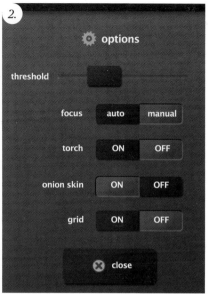

**1.**

Open the iMotion HD app and rotate the phone into a landscape position. This will ensure that the app records the frames in 720p and will therefore result in much better image quality.

In an ideal world, you should never touch a camera when you are taking a frame for an animation. Luckily, iMotion HD has a built-in system that uses sound to trigger the capture button. Choose "Mic" and then whistle or clap to see how far the sound meter goes, then slide the marker to sit at the maximum loudness. This means that when the microphone detects a sound as loud as the one you just created, it will take a frame.

**2.**

Tap "Start" to begin the animation and access some further settings. Go to "Options" and set the focus to "Manual." Then, scroll down and turn onion skinning to "On" and tap "Close."

**3.**

Place the iPhone in the tripod mount and place something in front of the camera to focus on. Then, tap and hold on the screen so that the camera locks its focus. This is important, as you don't want the camera shifting focus during the animation.

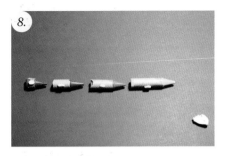

## 4.

Because the camera on the iPhone has no optical zoom, we need to mark out the edges of the frame so we know where items need to come in and out of the shot. The easiest way to do this is to place a pencil on the black poster board and point it to the corner of the screen on the iPhone. Then mark the position of the tip of the pencil on the poster board.

## 5.

Do the same for all four corners and then join all of the marks up with a ruler to mark out the frame of the camera's field of view.

## 6.

Turning your attention to the "set," use your chosen materials to decorate it. While Popsicle sticks and reflective craft butterflies aren't necessarily found at the bottom of the sea, they do make for a stylized seabed. Use reusable adhesive to stick down the decorative pieces, as well as using masking tape to secure the black poster board in place.

## 7.

Once you have everything in place, you can begin animating the various collected objects, in this case as if they were swimming underwater. The onion-skinning function in iMotion HD is extremely handy when animating objects on a 2D surface.

## 8.

For animating the crayons, I chose four of them and cut them incrementally smaller. I then animated these using "Substitute Animation," which is when an object is swapped out for a slightly different-shaped one between frames. This gave the appearance of the crayon alternately contracting and expanding to push itself through the water.

**9.**

Once you have finished animating, remove the iPhone from the tripod mount and double tap "Stop." iMotion HD will now play the animation back to you on a loop. From here you can change the playback speed and also edit out frames you don't want by going to Tools. You can also use the frame advance buttons and then press delete when you are on a frame you don't want.

**10.**

To export the animation and share your work with others, simply tap on "Export" and choose which method you want, such as sending directly to YouTube or the online animation "iMotion gallery" hosted by Fingerlab.

**11.**

Of course, you can also email yourself a copy of the movie file from the iPhone and then you can view your animation anywhere you wish.

# Section Three
# POSTPRODUCTION

Still from *The Rooster, the Crocodile, and the Night Sky* by Eimhin McNamara

*Editing and Creating Movies for Sharing*
# EDITING OVERVIEW

When creating a short movie or animation, there are three main stages to the process: pre-production, production, and postproduction. Postproduction usually involves taking everything you did during production (in this case, animating) and organizing it into a final, watchable movie that you can share with others.

The first step is to turn the image sequences you created during animation into digital movie files. Then edit these movie files together in order (using the storyboard for reference) with sound, titles, and credits. Adding music or sound effects to your animation can make it a lot more interesting to watch, just as putting different shots together can make the story in your animation more interesting than having one continuous shot.

Once you have completed the final edit of your animation, you can adjust the colors before rendering out a complete final movie file of your animation. You can upload the final movie file to websites such as YouTube or Vimeo to share with your friends, family, and the rest of the world.

## Operating systems

The type of operating system you use will depend on the type of computer you are using. If you work on a PC, you will most likely be using a Windows operating system, and if you work on an Apple computer, then you will be using OSX. Regardless of which operating system you use, there are editing packages available that will let you edit your final animation together and prepare it for uploading and sharing. Check out Chapter 2, pages 26–27, for a basic software guide for your operating system.

The next few pages will take you through the processes of turning an animation image sequence into a movie file, editing your movie files together with titles, and rendering a final movie file of your animation.

Each step is accompanied by an online tutorial movie that you can watch to learn more and see the process in action, using some of the animations from the tutorials in the book.

The key to postproduction is to understand that it is a creative process. Apart from a few basic principles about editing and rendering movie files, you can experiment as much as you like to help make your animations really come to life.

Windows operating system.

Apple OSX operating system.

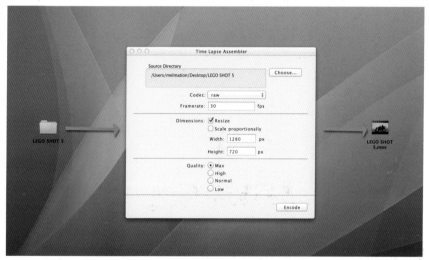
The basic process of turning an image sequence into a movie file.

Online tutorial movie from each section in this chapter.

Add effects to bring everything together.

# IMAGE SEQUENCE TO MOVIE FILE: OSX

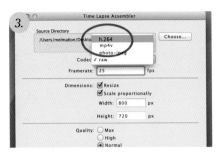

Whether using a digital stills camera or capture software such as Helium Frog, you will end up with a folder containing a sequence of images that you have shot for your animation. These images can't be played back like a movie file to see the animation, so you need to use a free piece of software called Time Lapse Assembler.

## The setup

Once you have downloaded and installed Time Lapse Assembler, double click the application icon to launch the software. You will see the main interface that gives you access to basic settings to create a movie file from an image sequence.

**1.**
At the top of the window, click "Choose…" to select the folder where the image sequence appears.

**2.**
Find the folder and click "Open."

**3.**
Click the drop-down list beside "Codec:" and choose the option "h.264."

**4.**
Change the "Framerate:" option to the frame rate you are working in. I am going to choose 30 fps.

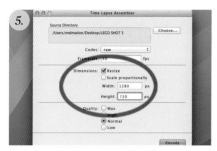

Scriptura

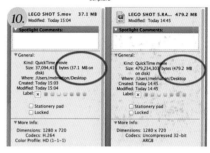

## 5.
To change the resolution of the movie file, uncheck the option "Scale proportionally" and type in the resolution of the movie you wish to create. I am choosing 1280 x 720, known as the resolution 720p.

## 6.
Change the "Quality" setting to "Max" and then click "Encode." Time Lapse Assembler will then ask where you want to save the movie file. Choose the location you want and click "Save."

## 7.
Time Lapse Assembler will then Encode (also known as Render) the movie file and save it in the location you chose in step 6.

## 8.
Go to the location where you told Time Lapse Assembler to save the movie file and open it to check that it has rendered correctly. This movie file is suitable for uploading to YouTube or playing on another computer. However, the codec used (h.264) removes a lot of visual information to reduce the file size. If you want to create a movie for editing, you will need to create a higher-quality movie file.

## 9.
Go back to Time Lapse Assembler and change the "Codec:" setting to "Raw" and click "Encode" to save the movie file.

## 10.
Right Click (or Ctrl+Click) both movie files to see the difference in file size. The movie rendered using the Raw Codec is bigger because it contains far more information, making it ideal for editing together with other shots using editing software.

# EDITING MOVIE FILES IN iMOVIE ON OSX

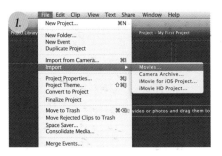

Scriptura

iMovie is a free movie-editing package included with every Apple computer. With iMovie you are able to import movie files, edit them, and add titles and basic effects that can greatly enhance your final animation. Using iMovie, you can do many things, but here are a few basics to get you started.

You will find the iMovie application in the applications folder on your Apple computer. Once you have opened the software, you will see the iMovie interface.

**1.**
Bring your animation shots into iMovie by going to File > Import > Movies at the top of the window, then navigate to the folder where your movie files are.

**2.**
Change the "Optimize video" option to "Full—Original Size," make sure the "Copy files" option is checked and click "Import." The movie files will now import into iMovie, which may take a while depending on how many there are.

**3.**
The animation shots will appear in the bottom window. Click and drag along a shot to select part or all of the shot and drag it into the top right-hand window. This starts to form the timeline for your animation. Continue to add the other shots in the order you want to build up your animation.

Drag the mouse pointer over the clips you have added to the timeline to "scrub" through the animation as you add in different clips. Press the spacebar to play through the animation.

**4.**
To add titles to your work, click the Text icon on the right hand side of the application, just below the viewer window. This will open the titles library.

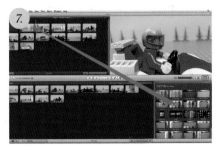

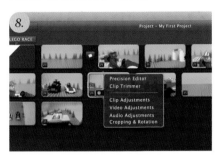

**5.**
Choose a title style and drag it to the start of the timeline.

**6.**
An option window will appear that lets you choose the style of background you want for the titles. Choose a style and then edit the text for the titles in the viewer window.

**7.**
To add transitions between your shots, click on the Transition icon at the right-hand side, below the viewer window. Add the transitions by clicking and dragging them between the shots in the timeline.

**8.**
Click on the Cog symbol of a clip when your mouse is over it and click on "Video Adjustments."

**9.**
Here, you will be able to adjust basic settings, such as exposure, brightness, contrast, and saturation.

# RENDERING MOVIE FILES FOR UPLOAD IN iMOVIE ON OSX

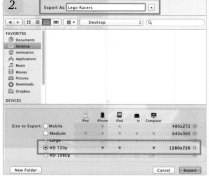

Continuing to use iMovie on OSX, we are now going to look at both rendering the edited animation into a movie file and also sharing it online. This process is quite straightforward, but it is important to get it right so that the final version of your animation will look good and play correctly.

**1.**
Open the iMovie project of your edited animation. At this point, it is a good idea to play through the entire movie to check that everything is how you want it.

**2.**
To render a movie file out that can be played on other computers that have QuickTime player installed, go to Share > Export Movie, which will bring up the export options dialogue box.

Give the movie a name and choose the size or resolution that you want the movie to be rendered out at. You will see that each option has its own optimized devices for playback. Every option, except Mobile, will play back on a computer that has QuickTime player installed.

### 3.
If you want to share your work widely, it is a good idea to upload the video to YouTube. Because YouTube is so popular, most computers will allow users to watch your finished animation online.

### 4.
To render and upload to YouTube directly from iMovie, go to Share > YouTube, which will bring up the YouTube publisher dialogue box.

### 5.
Click "Add" and enter your YouTube username (if you don't have one, sign up to YouTube to get one). Enter your YouTube password directly below and then choose a category for your animation to be listed under. Finally, give your animation a name and a description, both of which will appear on the YouTube website when it has uploaded.

### 6.
Make sure to type your keywords with a comma and a space between them to help YouTube recognize them. Choose the resolution of the movie file you want to render and upload and then uncheck "Make this movie personal" if you want everyone to be able to see your animation. Then click "Next."

### 7.
A terms of service window will appear, and if you agree to the terms, click "Publish." iMovie will now render and upload your movie directly to YouTube. Once it is finished, you can check out your own animation online.

# IMAGE SEQUENCE TO MOVIE FILE: WINDOWS

Whether using a digital stills camera or capture software such as Helium Frog, you will end up with a folder containing a sequence of images that you have shot. Because these images can't be played back like a movie file to see the animation, you need to use a free piece of software called VirtualDub.

VirtualDub can be downloaded for free from www.downloadvirtualdub.com and, unlike a traditional program, it is run from within a folder of files that you download.

**1.**
Open the VirtualDub folder that you have downloaded and double click the VirtualDub application icon to launch the software.

**2.**
With the application open, go to File > Open video file, which will then open the file navigation window. Go to the location of your image sequence, select the first image, and then select "Open."

Next, you need to set the frame rate for the video file. Do this by going to Video > Frame Rate, which will open the Video Frame Rate control window.

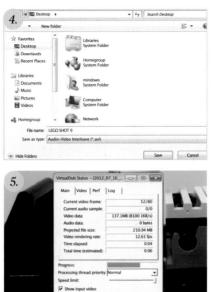

### 3.

Select the "Change frame rate to (fps)" option and enter the frame rate at which you want your rendered movie file to play. It is best to keep this the same as the frame rate at which you shot your animation.

### 4.

Click "OK" and then go to File > Save as AVI to save your movie file as an AVI.

### 5.

Give your file a name and choose a location for the movie file to be rendered to. Then click "Save" for VirtualDub to start rendering your movie file.

### 6.

Once the movie file has been rendered, you will be able to close the program and locate the file. It should open in Windows Media Player when double clicked.

Note that this file is uncompressed, meaning that VirtualDub has kept all the visual information in the rendered movie file. While it may not play back smoothly in Windows Media Player because of the file size, it is perfect for bringing into an editing package such as Adobe Premiere Elements to edit with.

The movie files that you export from VirtualDub will be quite large, so ensure you have enough room to save all your rendered animation shots before editing. Always be sure to save the image sequence in case you need to go back and render another movie from it.

# EDITING MOVIE FILES IN ADOBE PREMIERE ELEMENTS IN WINDOWS

Adobe Premiere Elements is an inexpensive editing package available for Windows-based computers. Premiere Elements is a stripped-down, simpler version of Premiere Pro, but although it costs a fraction of the price, it contains a lot of tools for editing and creating your final animation.

Once Premiere Elements is installed, locate the application icon in your start menu and double click to launch the software. You will then see a welcome screen. Click on "New Project" and give your project a name. Click "Next."

**1.**
To bring rendered animation shots *(see pages 160 and 161)* into Premiere Elements, go to File > Get Media From > Files and Folders, then navigate to the folder where the shots are stored.

**2.**
Once the shots have loaded, check the "Details" box to see the names of the shots in the media browser.

### 3.
Click and drag the first shot from the media window into the timeline along the bottom. This will start to build the timeline of your animation. Continue to do this with the other shots until you have all of them in the timeline, in the correct sequence.

### 4.
As you add shots, the timeline will expand. You can zoom in and out of the timeline using the timeline zoom bar underneath the viewer window. Drag the ends of the zoom bar to zoom in or out. Drag the playhead left or right to scrub through the timeline, and press the spacebar to start or stop playback.

### 5.
To add titles to your animation, click the "Edit" tab in the top right-hand window and then choose the "Titles" icon tab. Click and drag a title style into the timeline. Double click on the text to edit it and to change the font or color of the text.

### 6.
To add transitions between your shot, click on the "Transitions" icon under the "Edit" tab to bring up a selection of transitions. Click and drag any of the transitions onto the timeline, between the shots you want to transition. Scrub through the timeline to check if the transition works.

### 7.
To add basic effects, choose the "Effects" icon. Click and drag a filter onto a shot in the timeline and click "Apply."

### 8.
You can then click on "Edit Effects" to adjust the parameters of the effect. When you are done, click "Apply."

# RENDERING MOVIE FILES FOR UPLOAD IN PREMIERE ELEMENTS IN WINDOWS

Once you have edited your animation together in Premiere Elements, you can finish off the editing process by rendering a final movie and uploading it to a video-sharing site such as Vimeo. This process is quick and easy to do, but it must be done right so that your video will play back correctly online.

**1.**
Open the Premiere Elements project file for your animation. As always, play your final edit back to make sure that everything is how you want it.

**2.**
To render out a movie file that can be played back on other computers or uploaded to Vimeo, click the "Share" tab on the top right-hand window and then choose "Computer." This will bring up the export options dialogue text.

**3.**
Choose the "Windows Media" option, and then from the "Presets" drop-down menu choose "HD 720p 30," as this is the format in which the images were captured. Give the movie a name and choose where you want to save it, then click "Save." Premiere Elements will then render out your final animation as a Windows Media movie file.

## Uploading to Vimeo

Vimeo is a popular video-sharing site. Like YouTube, Vimeo will actually re-render your video after it has uploaded so that it can be played back easily to people who view the video online.

### 4.

To upload your animation to Vimeo, go to www.vimeo.com and create an account, then log in. Once you are logged in, go to "Upload a Video" on the right-hand side of the webpage.

### 5.

On the next page, click "Choose a Video to Upload" and then navigate to where you saved your final animation movie file and click "Open." Choose "Upload Selected Video." Vimeo will start to upload the video file. While it does so, you can fill in the details for your video.

### 6.

Once your video has uploaded you can click "Go to Video!" Depending on how busy Vimeo is, you may have to wait until your video has finished being re-rendered. Once it is ready, you will be able to view your video online along with the rest of the world.

# RESOURCES

## Books

**Adobe Premiere Elements 10 Classroom in a Book**
by Adobe Creative Team, Adobe, 2011

Further information on editing in Premiere Elements

**The Animator's Survival Kit** by Richard Williams, Faber and Faber, 2009

Reference for study of character movement within animation

**Collins Complete Photography Course**
by John Garrett and Graeme Harris, Collins, 2008

Comprehensive photography book

**Cracking Animation** by Peter Lord and Brian Sibley, Thames and Hudson, 2010

An excellent resource on the Stop-Motion Techniques used by Aardman Animations

**Frame by Frame Stop Motion: The Guide to Non-Traditional Animation Techniques** by Tom Gasek, Focal Press, 2011

Further information on Stop Motion Animation techniques

**How to Write for Animation**
by Jeffrey Scott, Overlook Press, 2004

Further information on how to create stories for animation

**The Illusion of Life: Disney Animation**
by Frank Thomas and Ollie Johnston, Hyperion, 1997

Believed to be the original source for the 12 Principles of Animation

**My iMovie** by Dave Caolo and Steve Sande, QUE, 2013

Further information on editing in iMovie

**Prepare to Board!** by Nancy Beiman, Focal Press, 2007

A fantastic resource on different pre-production techniques to help you develop a story for your animation

**Stop Motion: Craft Skills for Model Animation** by Susannah Shaw, Focal Press, 2008

Further information on craft skills for Stop Motion Animation

**The Storyboard Design Course: The Ultimate Guide for Artists, Directors, Producers and Scriptwriters**
by Guiseppe Cristiano, Thames and Hudson, 2008

Further information on creating storyboards

**Timing for Animation**
by Harold Whittaker and John Halas, Focal Press, 2009

Further information on timing for animation

# Websites

### 11 Second Club
www.11secondclub.com

Online monthly animation competition full of fresh inspiration work created by animators from around the world

### Animation World Network
www.awn.com

Community resource for all current animation news

### Animate Clay
www.animateclay.com

Forums and message-board community for all things Stop Motion

### Bricks in Motion
www.bricksinmotion.com

A forum-based community for creators of LEGO animations

### Cowboy with an Icecream
www.vimeo.com/channels/cowboyicecream

A fantastic collection of inspirational animations on Vimeo

### Skwigly
www.skwigly.co.uk

Animation magazine with animation news, interviews, and podcasts

### Stopmotionanimation.com
www.stopmotionanimation.com

An online collection of resources for the Stop Motion Animation community

### Stop Motion Magazine
www.stopmotionmagazine.com

An online quarterly publication devoted to the art of Stop Motion Animation

# Suppliers

### Amazon
www.amazon.com

Amazon can supply many of the tools and materials mentioned in this book, such as liquid latex, Perspex, plywood, clay, and Plasticine.

### Animation Supplies.net
www.animationsupplies.net

Animation equipment and materials for Stop Motion Animation, including Newplast Plasticine, rostrums, and ball-and-socket armatures

### Dragon Frame
www.dragonframe.com

Download a free trial copy of Dragon Frame for PC or MAC

### Helium Frog
www.heliumfrog.com

Download a free copy of Helium Frog for PC

### Julian Clark Studio
www.julianclarkstudios.com

Home of the AnimaSapien ball-and-socket armature and other parts created for Stop-Motion Puppet Animation

### LEGO
www.lego.com

LEGO supplies

### Stop Motion Works
www.stopmotionworks.com

List of suppliers for Stop-Motion materials and tools

### Van Aken Clay
www.vanaken.com

Information about Van Aken clay, including listed retailers

# GLOSSARY

**Adobe After Effects**
A video-compositing software produced by Adobe. This software makes it possible to layer pieces of video footage on top of one another, as well as to apply and control digital filters.

**Adobe Photoshop**
An image-editing software produced by Adobe. This software makes it possible to layer still images on top of one another, as well as to apply and control digital filters.

**Adobe Premiere Elements**
A video-editing software produced by Adobe. This package contains basic tool and filters for use when editing digital video footage.

**animatic**
A video depicting a sequence of storyboard pictures, played in order, each lasting a specific length of time and accompanied by sound.

**animation wire**
An alloy metal mainly consisting of aluminum, it is soft and easy to bend but will hold its shape when untouched.

**arcing**
Within animation, this relates to the movement of that which is being animated. An object moving in an arc reads as a much more natural motion than an object that changes and breaks its direction of movement.

**armature**
The structure inside an animation puppet, used to help it stand up straight and stay in position between shots.

**aspect ratio**
The ratio of the width to the height of an image on a screen.

**AVI**
Stands for Audio Visual Interleave. A movie-file format compatible with Windows computers.

**ball and socket**
A type of armature used for animation puppets. Metal balls rotate inside a socket, creating moving joints.

**Claymation**
An animation created using mostly clay or Plasticine as the medium.

**Codec**
A mathematical process used to create digital video information that can play back on a computer. Codec stands for "Compression—Decompression."

**compact digital camera**
A digital camera that is not part of a smartphone and cannot have its lens detached from the body. Usually small enough to be held and operated with one hand.

**compression**
A method of removing digital information to either reduce file size or create file compatibility.

**digital single lens reflex (DSLR) camera**
A digital camera that utilizes a mirror system between the back of the lens and the viewfinder, allowing the user to see "through the lens" when shooting. DSLR cameras have interchangeable lenses.

**down shooting**
When the camera is pointed directly downward while animating or filming.

**Dragon Frame**
A Stop-Motion capture software produced by DZED. This software can capture images from a digital camera attached to the computer running Dragon Frame. The software can then play back the images, as well as offer many tools to aid an animator, such as onion skinning.

**Fimo**
A clay-like substance that can be sculpted and shaped before it is baked in an oven and turns solid. After baking, Fimo can be sanded, drilled, and painted.

**Fingerlab iMotion HD**
An application available for smartphones produced by Fingerlab. This software allows you to capture images from the smartphone's camera and play them back in sequence, as well as offering useful tools such as onion skinning.

**frames per second (FPS)**
Also known as frame rate. Relates to how many images are played back every second to the viewer, in order to create the illusion of motion.

**Helium Frog**
Free software available to download for PCs. Helium Frog allows you to capture images from a webcam and then play them back in sequence, as well as offering useful tools such as onion skinning.

**high definition**
Relating to the resolution of an image or video footage. Full high-definition images and video have a resolution of 1920x1080.

**horizontal shooting**
When the camera is pointed horizontally while animating or filming.

**image sequence**
Relates to a computer folder containing image files that are the sequential frames of an animation or video.

**iMovie**
Video-editing software produced by Apple and available for Macs. This software contains basic tools and filters for use when editing digital video footage.

**infinity curve**
A backdrop used for photography where the material being used comes from a wall becoming the floor through a gradual curve, making the backdrop seamless.

**iStopMotion**
Stop-Motion capture software produced by Boinx and available for Macs. This software can capture images from a digital camera attached to the same computer. The software can then play back the images and offers many tools to aid an animator, such as onion skinning.

**light box**
A box consisting of a transparent or semi-opaque lid with a light that shines through from inside the box. Within animation, this can be used for both 2D Drawn Animation and Silhouette Animation.

**Monkey Jam**
Free software available to download for PCs. It allows you to capture images from a webcam and then play them back in sequence, as well as offering basic editing tools for editing your animation as you go.

**onion skinning**
A term used to describe being able to see the previous or next image of an animation as well as the current image.

**pixel**
The smallest measurable part of a digital image. A pixel can refer to both a square of color within a digital image or a single illuminated point on a computer screen or television.

**pixilation**
When subject matters that can move of their own accord (such as humans) are animated using stop-frame techniques.

**rendering**
An action carried out by a piece of software running on a computer to produce a new piece of information from many other pieces of information. Within animation and video editing, this task takes place when a sequence of images, or a sequence of video files, are rendered into one single video file.

**resolution**
Relates to the size of an image, usually measured in pixels.

**rostrum**
A specialist piece of apparatus used for holding a camera above the subject matter, putting the camera into a down-shooting position.

**shooting blind**
Animating without the aid of onion skinning or the ability to play back the animation as you are shooting.

**shooting on 2's**
When two frames are captured for each movement of the medium being animated. The number of frames captured per movement can also be referred to as shooting on 1's and shooting on 3's.

**Sony Vegas**
A video-compositing software produced by Sony. This software makes it possible to layer pieces of video footage on top of one another, as well as to apply and control digital filters.

**storyboard**
A collection of images on one page, or "board," depicting a sequence of shots for a film or animation.

**"T" pose**
The shape created by an animation puppet's pose when it is being created.

**tie-down**
A method of connecting a puppet or prop to the surface on which it is being animated. Pins and bolts are two common methods of tie-down.

**Time Lapse Assembler**
A free piece of software available to download for Macs. This software allows the user to turn a sequence of images into a movie file.

**VirtualDub**
A free piece of software available to download for PCs. Among many other uses, this software allows the user to turn a sequence of images into a movie file.

**webcam**
A small digital camera with a microphone, which is either built into a computer monitor frame or is located on top of a monitor, attached to the computer via a USB port. A webcam streams video information to the computer.

**widescreen**
Relates to the shape of video footage, images, computer monitors, or televisions. Widescreen usually relates to the image's aspect ratio of 16:9, which is visually wider than the squarer-shaped 4:3 aspect ratio.

# INDEX

Aardman Animations 56, 62, 138

action figures 74

Adobe 26-27, 86, 88, 96, 100, 112, 161-165

After Effects 96, 100, 112

Afternoon 58-59

Aladdin 106

aluminum wire 37

Angels We Have Heard on High 86-87

animation wire 35, 37, 68, 70, 76-78

Animator HD 96

Apple 26-27, 88, 136, 142, 146-149, 156-159

apps 136, 142, 144-147

arcing 13

The Arctic Circle 70-71

armatures 37, 68-69, 72, 76-78

articulation 102, 107

aspect ratio 29, 33

At The Opera 62-63

attachments 18-19, 136

available light 20

baby wipes 65

background 95-96, 116-117, 120-121, 138, 157

Baladi, Camille 100-101

ball and socket 68, 72

baseplates 108-109, 117-118, 120-121

8-Bit Trip 109, 114-115

Blackton, J. Stuart 6, 84

blinking 82

Blu 88

blueprints 76

Boivin, Patrick 74-75

bolt tie-down 69

boots 70

bouncing light 23

Brick Films 108-109

C-clamps 35

The California Raisins 60

camera angles 47, 120

camera supports 18-19

camera types 17

Canon RemoteCapture 58, 126

capture software 24-25, 28-29, 31, 154, 160

cell phone animation overview 136-137

cell phones 16-18, 74, 133, 138, 140, 144

chalkboard animation overview 84-85

clay types 56

claymation overview 56-57

clip light 21

close-up shots (CU) 47, 112, 114, 121

clothes 68, 78

clouds 83

codec 154-155

Collide Entertainment 86

color 20, 22, 31, 56, 67, 72, 85, 122, 130, 132, 145, 152, 163

compact digital cameras (CDCs) 15, 17, 88, 90

Computer-Generated Imagery (CGI) 112, 114

Coraline 68

Corel 126

cross fades 53

cutout animation overview 94-95

3D Studio Max 112, 142

desk lamps 21, 30, 134

detailed sets 38, 40

detailed storyboards 50

dialogue 52

diffused light 22, 80, 134

digital cameras 14-17, 106, 110, 136, 154, 160

digital single lens reflex cameras (DSLRs) 15-17, 25, 28, 30, 58, 60, 72, 74, 80, 96, 100, 106, 114, 124, 126, 128, 134, 142

direct light 22

direction 51

Dot 138-140

down shooting 19, 30, 64, 67, 106, 133, 137, 145-147

Dragon Frame 24-25, 30, 33, 60, 80-81, 83, 86, 106-107, 134-135

Dragon Stop Motion 72, 74

drawing tips 50

Driessen, Paul 62

dry-erase markers 84, 92-93

duct tape 34, 81, 90, 105-106

Dumala, Piotr 58

Earthquake Putty 112
easels 90, 92
easing in/out 11
eco bulbs 21, 104
editing/editing software 26-27, 29,
    53, 67, 86, 100, 110, 124, 126,
    152-153, 155-156, 162-163
epoxy glue 36
eyes 68, 70, 78, 82

The Family Bones 128
field of view 148
file size 155, 158, 161
fill light 23
filming 52, 80
Fimo 35, 37
Final Cut Pro 27, 124, 128
Fingerlab iMotion HD 25, 136,
    142, 146-149
Flash 96
flicker 112, 114, 119, 145
flip pads 6, 10, 142
fluorescent light 21, 30, 80, 83, 86,
    110, 114, 120, 134
foam 37, 68-69, 72, 76-79, 142
focus 31, 33, 67, 81, 107, 112,
    119-120, 147
Foster, Rebecca 88-89
frame rate 29, 32-33, 53, 154,
    160-161
Frame Thief 62
Fry, Arthur 124
full high definition (full HD) 28-29

gels 22, 72
graph paper 132-134
Grist, Hilary 86-87
Gulp 140-141

halogen lamps 21, 72
Helium Frog 24-25, 30, 32, 66,
    92-93, 120, 154, 160
Hickox, Michael 110-111
high definition (HD) 28-29
hinged puppets 94, 103
Hirst, Damien 88
horizontal shooting 30
Horowitz, Kevin 112-113
hot glue gun 34
hot sets 40
The Hours 88
Humorous Phases of Funny Faces
    6, 84

If You Ever Need Someone 128-129
Image Flux 128
iMovie 26-27, 88, 156-159
impact glue 36, 76, 78
infinity curve 39, 117
iPhone 136, 142, 146-149
Iron Man vs. Bruce Lee 74-75
iSight 25
iStopMotion 24-25, 98
Ivey, Javan 98-99

jitter 18
Johnny and June 88-89
Johnston, Ollie 11

Kaminar, Aaron 128-129
keywords 159
Kineograph 6
King Kong 7
Kinget, Antoine 72-73
Kontramax 142-143
Kucia, Jerzy 58

Larson, Gary 86
layout 48-49
Le Dantic, Marine 100
LEGO animation overview 108-109
LEGO Mini Golf 110-111
LEGO set building 116-117
Lenica, Jan 58
lenses 16, 114, 124, 136, 145
Lewin, Angie 140
light 19-23, 40, 50, 57-58, 66-67,
    72, 80-81, 83, 88, 90, 92, 100,
    104, 110, 112, 114, 118-120,
    134-135, 140, 142, 145-146
light boxes 94, 104-107
Linnett, John Barnes 6
liquid latex 36, 76, 78-79
Live View 15, 96

3M Canada 124
McNamara, Eimhin 96-97
magnetic tie-down 69
materials 36-37, 68-69, 95, 146
medium (M) shots 47
mid shots 40
mini-figures 108, 110, 112, 116, 120-121
The Mirror Cube 142-143
Monkey Jam 25, 112
montage 95
mouths 82
multi-function sets 40
Muybridge, Eadweard 6
My Paper Mind 98-99

needle-nose pliers 35, 82
Neodymium magnets 69
Newplast Plasticine 56
newspapers 81
Noce, Kim 56
Nokia 138, 140

Ode to a Post-It Note 124-125
oil-based clay 56, 58
onion skinning 15, 25, 31, 60, 67, 83, 90-93, 134-136
OSX editing 153-159

panning 18, 51
paper 44, 102-103, 122
Parisse, Rémi 72-73
Park, Nick 7
Parry, Kevin 70-71
Patterson, Ed 138-141
Persistence of Vision theory 10
photos 51, 124, 136

Photoshop 88
Pigeon Pilfer 60-61
pin tie-downs 69, 81-82
pixilation 6-7, 62, 86, 100, 136, 146
Plasticine 35, 56, 67
Plucinska, Izabela 58-59
Plympton, Bill 62
plywood 40
Des Pop et des Up 100-101
poses 79, 102
Post-it Note animation 122-123
Post-it Note types 122, 130-131
Post-it Stop-Motion Thank You 126-127
poster board panels 133
postproduction 27, 70, 124, 128, 142, 150-165
Pour une Pépite de Plus 72-73
pre-production 124, 152
Premiere 26-27, 86, 100, 161-165
props 42-45, 81-82, 110, 116-118
puppet overview 68-69
puppet-making 76-79, 94-95, 102-103

QuickTime 158

recording 52
Redigh, Tomas 114-115
Reiniger, Lotte 7, 94
rendering 153, 155, 158-159, 161-162, 164-165
replacement object animation 62
resolution 14-15, 17, 28-29, 32, 136, 155
reusable adhesive method 103
Roi, Arnaud 100-101

The Rooster, the Crocodile, and the Night Sky 96-97
rostrums 19, 30, 57, 64, 66
Rybczynki, Zbigniew 58
Rymdreglage 109

scale 60, 116-117, 138, 155
scalpels 34, 135
Schulz, Charles 86
Schwartz, Jacob 112-113
sculpting tools 35
Sesame Street 98
sets 38-41, 70, 80, 82, 86, 148
settings 31, 119
setups 30-33, 57, 64-67, 80-83, 90-93, 104-106, 109, 116-119, 133-135, 138, 144-149, 154-155
sewing method 103
shooting 80-83, 90-93, 106-107, 120-121, 131, 134-135
shooting on 1's/2's 32, 81
shooting blind 15, 24, 31, 66, 90
shot length 49, 52-53
silhouette animation overview 94-95
sketching 132
skies 117, 120
smartphones see cell phones
software 24-29, 31, 53, 58, 62, 74, 90, 92, 96, 100, 114, 118, 120, 134, 136, 144, 146, 153, 155, 160
Sony Vegas 27, 110, 142
Southworth, Mike 86-87
split-pin method 103
staging 12, 83
standing up 69
Stearns, Jeff Chiba 124-125
stepladders 131

Stevenson, Michael 60–61
Stopmotion Pro 25
stopwatches 52
storyboards 46–53, 58, 81, 86, 100, 118, 120, 128, 132, 152
Studd, Will 138–141
Sturton, James 126–127
Sumo Science 140
Svankmajer, Jan 7, 62
Szczechura, Daniel 58

telephoto lenses 16
themes 116
Thomas, Frank 11
Time Lapse Assembler 154–155
timing 13, 31, 52–53
Tirard, Nicolas 100
titles 152–153, 156–157, 163
toasters 44
Tomorrow Is a Chance to Start Over 86
tools 34–35
trees 43, 120–121
tripods 18–19, 30–31, 40, 57, 64, 80, 90, 92, 119–120, 137, 142, 144, 146–147

umbrellas 44–45, 82–83

Van Aiken clay 60
Video Games with James 126
VideoStudio Pro X4 126
Vimeo 152, 164–165
VirtualDub 160–161

Wallace and Gromit 60, 68
walls 40, 81–82
water-based throwing clay 56
Watterson, Bill 86
Webb, Justin 128
webcams 14, 17, 25, 28, 30, 32, 60, 66–67, 92, 106, 112, 117–120, 133
white balance 31, 119–120, 145
whiteboard animation overview 84–85
wide (W) shots 47
wide-angle lenses 16, 124
wide-screen 29
wind 83
window shooting 64
Windows editing 153, 160–165
Windows Movie Maker 27
Writer's Block 112–113

Yellow Sticky Notes: Canadian Anijam 124
YouTube 86, 110, 114, 126, 149, 152, 155, 159

Zaramella, Juan Pablo 62–63
Zoetrope 6
zooming 51, 80, 90, 93, 114

# CONTRIBUTORS

**Camille Baladi & Arnaud Roi, Up, Up Up**
www.upupup3d.com
Pages 100–101

**Patrick Boivin**
www.youtube.com/patrickboivin
Pages 74–75

**Bob Blunden**
limbic.webs.com
Page 68

**Jeff Chiba Stearns, Meditating Bunny Studio, Inc.**
www.meditatingbunny.com
Pages 124–125

**Joe Clokey, Prema Toy Co.,/ Premavision, Inc.**
www.gumby.com
Page 56, Image 1
© GUMBY and Gumby characters are registered trademarks of Prema Toy Co., Inc. All rights reserved.

**Pádraig Fagan**
www.vimeo.com/user729186
Pages 96–97
*The Rooster, the Crocodile and the Night Sky*, directed by Pádraig Fagan. Funded by The Irish Film Board, RTE, and The Arts Council. Produced by Barry O'Donoghue at Barley Films.

**Rebecca Foster**
www.beccafoster.co.uk
Pages 88–89

**Hilary Grist & Mike Southworth, Collide Entertainment**
www.collideentertainment.com
Pages 86–87

**Anna Harding, Aardman Animations**
www.aardman.com
Pages 138–141

**Michael Hickox**
www.youtube.com/mLchaelhLckoxfilms
Pages 110–111

**Javan Ivey**
www.javanivey.com
Pages 98–99

**Aaron Kaminar, Image Flux**
www.image-flux.com
Pages 128–129

**Antoine Kinget**
www.vimeo.com/antoinekinget
Pages 72–73

**Antoine Kinget and Rémi Parisse**
www.pourunepepitedeplus.com
Pages 72–73

**Kontramax**
www.kontramax.com
Pages 142–143

 **Dave McCall, BFI Archives department**
www.bfi.org.uk/archive-collections
Page 94, Image 3

 **Eimhin McNamara**
www.eimhinssecretblog.blogspot.ie
Pages 96–97

 **Kim Noce, Mew Lab**
www.mewlab.com
Page 56, Image 2

 **Rémi Parisse**
www.remiparisse.blogspot.com
Pages 72–73

 **Kevin Parry**
www.kevinbparry.com
Pages 70–71

 **Ed Patterson and Will Studd, Sumo Science**
www.sumoscience.com
Pages 138–141

 **Izabela Plucinska**
www.izaplucinska.com
Pages 58–59

 **Tomas Redigh**
www.rymdreglage.se
Pages 114–115

 **Lucy Roberts**
www.madebylucy.com
Page 66

 **Jacob Schwartz and Kevin Horowitz**
www.youtube.com/user/
QuigiboProductions
Pages 112–113

 **Michael Stevenson**
www.mstevenson.net
Pages 60–61

 **James Sturton**
www.youtube.com/chipsturs
Pages 126–127

 **Juan Pablo Zaramella**
www.zaramella.com.ar
Pages 62–63

# ACKNOWLEDGMENTS

My thanks to the team at RotoVision: Isheeta Mustafi, Jacqueline Ford, Jane Roe, Cath Senker, and Lucy Smith. It has been a privilege to work with you all.

Thanks to Aimee, my long-suffering fiancée/animation widow, for being my third hand and keeping me sane throughout this project and all the others; Deirdre O'Connell for being a true friend and source of support; Kim Noce and Shaun Clark for helping me believe that this book was something I could do; Alice Savage for being my research scout in Germany; Scott Bevan for supplying handsets; my mother and father for understanding that animation is a real job; my little sister, Sara, for making sure things were "OK with the book"; the animation students and Greg Boulton at Sheffield Hallam University, who I have the great fortune to teach and work with.

I would also like to thank the amazing artists and animators who agreed to be in this book. Without hesitation, they agreed to share their work and answer my many questions about their projects to help this book become what it is.

Finally, I would like to thank you, the reader, for taking the time to read this book. I hope you have found some inspiration among the pages. It has been a privilege to write them for you.

Thank you

Melvyn Ternan